# Technologien für die intelligente Automation

## Technologies for Intelligent Automation

## Band 11

**Reihe herausgegeben von**
inIT – Institut für industrielle Informationstechnik
Lemgo, Deutschland

Ziel der Buchreihe ist die Publikation neuer Ansätze in der Automation auf wissenschaftlichem Niveau, Themen, die heute und in Zukunft entscheidend sind, für die deutsche und internationale Industrie und Forschung. Initiativen wie Industrie 4.0, Industrial Internet oder Cyber-physical Systems machen dies deutlich. Die Anwendbarkeit und der industrielle Nutzen als durchgehendes Leitmotiv der Veröffentlichungen stehen dabei im Vordergrund. Durch diese Verankerung in der Praxis wird sowohl die Verständlichkeit als auch die Relevanz der Beiträge für die Industrie und für die angewandte Forschung gesichert. Diese Buchreihe möchte Lesern eine Orientierung für die neuen Technologien und deren Anwendungen geben und so zur erfolgreichen Umsetzung der Initiativen beitragen.

Weitere Bände in der Reihe http://www.springer.com/series/13886

Jürgen Beyerer · Alexander Maier ·
Oliver Niggemann
Editors

# Machine Learning for Cyber Physical Systems

Selected papers from the International
Conference ML4CPS 2017

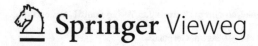 Springer Vieweg

*Editors*
Jürgen Beyerer
Institut für Optronik, Systemtechnik
und Bildauswertung
Fraunhofer
Karlsruhe, Germany

Alexander Maier
Industrial Automation
Fraunhofer-Anwendungszentrum
Lemgo, Germany

Oliver Niggemann
inIT – Institut für industrielle
Informationstechnik
Hochschule Ostwestfalen-Lippe
Lemgo, Germany

ISSN 2522-8579          ISSN 2522-8587   (electronic)
Technologien für die intelligente Automation
ISBN 978-3-662-59083-6      ISBN 978-3-662-59084-3   (eBook)
https://doi.org/10.1007/978-3-662-59084-3

Springer Vieweg

This Springer Vieweg imprint is published by the registered company Springer-Verlag GmbH, DE part of Springer Nature
The registered company address is: Heidelberger Platz 3, 14197 Berlin, Germany

# Preface

Cyber Physical Systems are characterized by their ability to adapt and to learn. They analyze their environment, learn patterns, and they are able to generate predictions. Typical applications are condition monitoring, predictive maintenance, image processing and diagnosis. Machine Learning is the key technology for these developments.

The third conference on Machine Learning for Cyber-Physical-Systems and Industry 4.0 - ML4CPS - was held at the Fraunhofer IOSB in Lemgo, on September $25^{th}$ - $26^{th}$ 2017. The aim of the conference is to provide a forum to present new approaches, discuss experiences and to develop visions in the area of data analysis for cyber-physical systems. This book provides the proceedings of selected contributions presented at the ML4CPS 2017.

The editors would like to thank all contributors that led to a pleasant and rewarding conference. Additionally, the editors would like to thank all reviewers for sharing their time and expertise with the authors. It is hoped that these proceedings will form a valuable addition to the scientific and developmental knowledge in the research fields of machine learning, information fusion, system technologies and industry 4.0.

*Prof. Dr.-Ing. Jürgen Beyerer*
*Dr. Alexander Maier*
*Prof. Dr. Oliver Niggemann*

# Table of Contents

**Page**

Prescriptive Maintenance of CPPS by Integrating Multi-modal Data
with Dynamic Bayesian Networks ............................... 1
    *Fazel Ansari, Robert Glawar, Wilfried Sihn*

Evaluation of Deep Autoencoders for Prediction of Adjustment Points
in the Mass Production of Sensors ......................... 9
    *Martin Lachmann, Tilman Stark, Martin Golz, Eberhard Manske*

Differential Evolution in Production Process Optimization of Cyber
Physical Systems ...................................... 17
    *Katharina Giese, Jens Eickmeyer, Oliver Niggemann*

Machine Learning for Process-X: A Taxonomy ..................... 25
    *Felix Reinhart, Sebastian von Enzberg, Arno Kühn, Roman Dumitrescu*

Intelligent edge processing .............................. 35
    *Ljiljana Stojanovic*

Learned Abstraction: Knowledge Based Concept Learning for Cyber
Physical Systems ...................................... 43
    *Andreas Bunte, Peng Li, Oliver Niggemann*

Semi-supervised Case-based Reasoning Approach to Alarm Flood Analysis . 53
    *Marta Fullen, Peter Schüller, Oliver Niggemann*

Verstehen von Maschinenverhalten mit Hilfe von Machine Learning ...... 63
    *Heinrich Warkentin, Meike Wocken, Alexander Maier*

Adaptable Realization of Industrial Analytics Functions on Edge-Devices
using Reconfigurable Architectures ......................... 73
    *Carlos Paiz Gatica, Marco Platzner*

The Acoustic Test System for Transmissions in the VW Group .......... 81
    *Thomas Lewien, Ivan Slimak, Pyare Püschel*

# Prescriptive Maintenance of CPPS by Integrating Multimodal Data with Dynamic Bayesian Networks

Fazel Ansari[1,2*], Robert Glawar[1,2], Wilfried Sihn[1,2]

[1]Vienna University of Technology (TU Wien), Institute of Management Science, Research Group of Smart & Knowledge-Based Maintenance
[2]Fraunhofer Austria Research GmbH, Division of Production and Logistics Management, Vienna, Austria
*fazel.ansari@tuwien.ac.at

**Keywords:** Maintenance, CPPS, Prescriptive Analytics, Cause-Effect Analysis, Data Model, Bayesian Network.

**Abstract.** The complexity and data-driven characteristics of Cyber Physical Production Systems (CPPS) impose new requirements on maintenance strategies and models. Maintenance in the era of Industry 4.0 should, therefore, advances prediction, adaptation and optimization capabilities in horizontally and vertically integrated CPPS environment. This paper contributes to the literature on knowledge-based maintenance by providing a new model of prescriptive maintenance, which should ultimately answer the two key questions of "what will happen, when? and "how should it happen?", in addition to "what happened?" and "why did it happen?". In this context, we intend to go beyond the scope of the research project "Maintenance 4.0" by i) proposing a data-model considering multimodalities and structural heterogeneities of maintenance records, and ii) providing a methodology for integrating the data-model with Dynamic Bayesian Network (DBN) for the purpose of learning cause-effect relations, predicting future events, and providing prescriptions for improving maintenance planning.

## 1 Introduction

### 1.1 Key Terms of Knowledge-Based Maintenance

The emergence of cyber physical production systems (CPPS) and the transition to Industry 4.0 trigger a paradigm shift from descriptive to predictive and prescriptive maintenance, colloquially known as the highest maturity level of knowledge-based maintenance (KBM). Figure 1 elaborates on the four maturity and complexity levels of KBM. Notably, maintenance professionals and researchers use different terms to describe the four maturity levels of KBM, which in some cases remain either vague (not precisely determined or distinguished) or ambiguous (have two or more interpretations), such as the concepts of smart maintenance and maintenance 4.0. In this paper, we use the overarching term of KBM that refers to multiple concepts, which can be distinguished from one another (Cf. Figure 1). In particular, prescriptive maintenance involves modeling expert knowledge, machine learning, predictive data analytics and semantic reasoning to enhance and automatize decision-making processes by optimal selection and proposing the right strategies, tactics and action plans for foreseeing and handling problems pertained to the entire maintenance management. Prescriptive

J. Beyerer et al. (Eds.), *Machine Learning for Cyber Physical Systems*, Technologien für die intelligente Automation 11, https://doi.org/10.1007/978-3-662-59084-3_1

maintenance of CPPS combines descriptive, diagnostic and predictive analytics to not only understand and reason out past events, but also to anticipate the likelihood of future events and potential effects of each decision alternative on the physical space and associated business processes.

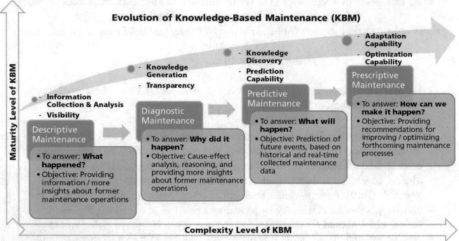

**Fig. 1.** Four Maturity Levels of KBM – From Descriptive to Prescriptive Maintenance

## 1.2    State-of-the-Art – Knowledge-Driven and Integrated Approaches

The state-of-the-art literature review on KBM reveals different approaches, which can be categorized into one of two general groups: i) *Knowledge-Driven Approaches* i.e. set of data-mining, machine learning and knowledge representation methods, which are evaluated in the context of diagnostic, predictive or prescriptive maintenance, including maintenance planning, monitoring and controlling activities, and ii) *Integrated Knowledge-Intensive Approaches or Reference Models*, including assistance systems, which aim at combining multiple data-sources and knowledge assets towards enhancing maintenance system intelligence. An overview of selected novel approaches is presented in this section.

Azadeh et. al. (2016) proposed an integrated multi-criteria approach, i.e. AHP-TOPSIS, to consider various parameters and non-linear characteristics of maintenance planning and ultimately optimize maintenance activities [1]. Windmann et. al. (2016) employed the Hidden Markov models and clustering-based condition monitoring for "comparing the actual process behavior with the behavior as predicted from the process models" especially for strictly continuous systems [2]. Moreover, Li and Niggemann (2016) proposed an approach for improving cluster-based anomaly detection on the process data to deduce the health status of systems and automatically detect anomalous statuses [3]. Paschke et. al. (2016) developed a generic data-driven model for active condition monitoring of automation systems, in particular for detecting wear-out or faulty situations "in machine subsystems, each consisting of a servomotor driving different parts of the machine" [4]. Ansari et. al. (2014), Dienst et. al. (2015) and Ansari and Fathi (2016) firstly proposed and then advanced an integrated concept for applying different meta-analytic and machine learning approaches for predictive maintenance of manufacturing systems, respectively, in [5], [6], and [7]. The integrated approach deals with various data structures (e.g. text and numeric data) as well as different aspects of maintenance (i.e. fault detection and cost controlling). Besides, Glawar et. al. (2016) outlined a holistic and anticipatory framework, which "enables the identification of maintenance-critical conditions and the prediction of failure moments and quality deviations" [8]. Moreover, Flatt et. al. (2015) proposed a maintenance assistance system, i.e. "an ontology-based context-aware framework which aggregates and processes data from different sources", using augmented reality and indoor localization technologies [9]. Wang et. al. (2017) discussed "a cloud-based paradigm of predictive maintenance based on mobile agent to enable timely information acquisition, sharing and utilization for improved accuracy and reliability in fault diagnosis, remaining service life prediction, and maintenance scheduling" [10]. The concept of "knowledge as a service" has been introduced by Abramovici et. al. (2017) that supports knowledge allocation and the recommendation of possible solutions in accordance with failure causes and similarity degree between former failure descriptions stored in the semantic knowledgebase [11]. Last but not least, Bärenfänger-Wojciechowski et. al. (2017) presented a reference integrated management approach, named smart maintenance, which combines key maintenance knowledge assets, namely humans, sensors, data-management and assistance-systems [12].

## 1.3    Research Project of Maintenance 4.0 and Beyond

In the research project Maintenance 4.0 an approach has been developed, which is based on the systematic exploration of three production-related perspectives, namely, product quality, process and machine. The machine is segmented in its maintenance relevant components in order to describe the behavior of the system. Load profiles can be calculated analytically based on machine data, such as acceleration data, positioning of tool slide, power consumption, and temperature sensors. Failure effects as well as quality-relevant cause-and-effect coherences are derived from maintenance relevant historical data collected from various data sources. Based on the captured information, a reaction model, which is supported by condition-based fault diagnostics, is able to forecast and anticipate failure moments. This model provides a set of rules, which is tailored to maintenance measures. Moreover, different data analysis methods such as statistical correlation analysis, association rule learning, and Weibull analysis are employed to discover the relations between variables, derive the relevant rules, and validate the model. The generated knowledge about maintenance and product quality trends and related effects is, therefore, interlinked with data from production planning. Providing a holistic view on all three perspectives, the Maintenance 4.0 model is able to propose quality and maintenance measures anticipatively. A suitable set of rules can be used as a basis to evaluate decision alternatives. Thus, it is possible to anticipate failure effects, to visualize quality trends, and to predict the wear-out for relevant machine components as well as the production system. On the top of this, the plant availability and the maintenance cost can be optimized, because maintenance measures are carried out in time and coordinated with the whole production system [8].

This paper aims to go beyond the scope of the Maintenance 4.0 project. In particular, we outline a generic model for prescriptive maintenance of CPPS considering multimodality and structural heterogeneity of maintenance records (Cf. Section 2). Using Event-Cost Schema, we integrate multimodal and multidimensional data collected from operational and management data flow of production systems. Furthermore, we concentrate on learning a dynamic model from structured data using stationary Dynamic Bayesian Networks (DBNs) to support the prediction of failure events and determination of potential effects on the quality of production planning processes as well as (consequential) maintenance costs (Cf. Section 3). The entire approach supports the provision of prescriptive recommendations for improving the forthcoming maintenance plan.

## 2    Prescriptive Maintenance Model (PriMa)

Figure 2 reveals the conception and building components of the Prescriptive Maintenance Model (PriMa). The underlying data warehouse continuously collects management (including cost data) and operation related data from three dimensions namely, machines, products and processes. These three dimensions are comparable to horizontal and vertical data flow of CPPS and associated processes, including actors and information systems. While the horizontal perspective refers to data collected either from operation or from management, the vertical perspective considers interlinking of operation and management data.

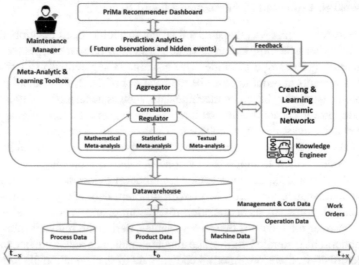

**Fig. 2.** Conception of PriMa

The maintenance records represent two major characteristics i) *Multimodality*, i.e. each signal or record provides independent information about different aspects of maintenance, e.g. while machine failure signal can reflect malfunction of a subsystem, it can also indicate inappropriate planning, monitoring or controlling which causes subsystem degradation and affects its remaining useful lifetime, and ii) *Structural hetero-*

*geneity*, i.e. maintenance records are constructed in two variations of 1) *structured records*, e.g. sensor and environmental information captured either by condition-based monitoring systems or via direct queries from machine programmable logic controller (PLC), or 2) *unstructured records*, e.g. maintenance text reports, audio and/or images collected by means of microphones or camera, respectively.

For the purpose of modeling and interlinking the aforementioned data, the Event-Cost Schema is developed using Data Vault 2.0 [13]. The Data Vault 2.0 facilitates building a data warehouse dealing with the multiple dimensions of scalability, in particular complexity due to volume, variety, velocity and veracity of data, and also accessing to integrated data [13]. Figure 3 depicts the Event-Cost Schema consisting four Hubs, which separate functional areas of maintenance of CPPS, namely; i) maintenance organization (MO), ii) production planning (PP), iii) cost controlling (CC), and iv) event tracking (ET). Hub-MO represents the semantic structures and related entities of maintenance activities, including ID number, type, category and timestamp. Hub-PP establishes the relations between various data sources used for production planning, including material and resource requirements as well as production planning processes. Hub-CC represents hierarchical cost relations, including infrastructure, logistics, personnel, and external costs associated with maintenance activities. Finally, Hub-ET presents the event-sensor(-network)-system relations on the shop floor, including type of event, which is determined through processing sensor (sensor-network) signals in conjoined with analysis of (sub-)system states. The relationships between the Hubs are stored in the Link-MO-PP-CC-ET.

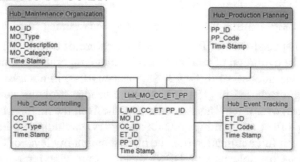

**Fig. 3.** Meta-model of Event-Cost Schema

Using the detailed data model, we semantically structure the key areas of maintenance and related data sources, and are able to associatively track maintenance costs-events in relation to production plans and maintenance activities. Furthermore, the meta-analysis is used to identify the strength of the relation between cost and operation parameters of CPPS (Cf. Figure 2). The cost parameters are defined as planned, unplanned and total cost (Cf. [7]) and the operation parameters are based on maintainability, availability and reliability factors (Cf. [6]). Meta-analysis algorithms are used for various types of data structures, i.e. structured and unstructured maintenance records. The meta-analytic component of PriMa consists of statistical, mathematical and textual meta-analytic algorithms, which have been discussed and evaluated in the earlier publications [5], [7]. The outcome of various meta-analysis algorithms should be correlated to exclude processing errors, to identify the interrelations, to aggregate the findings, i.e. provision of evidences for causality detection and finding patterns, and finally to recommend appropriate actions (decision alternatives). To learn from the instances of meta-analysis and to predict (future observations and hidden states), we may employ different statistical and machine learning approaches. In this paper, we exclude the structural heterogeneity, which has been discussed in [5] and instead place the focus on learning dynamic model from multimodal data using DBNs (Cf. Section 3).

## 3    Methodology for Integrating Multimodal Data with Dynamic Bayesian Network

A Bayesian Network (BN) "is a directed acyclic graph [...] that is composed of three basic elements", namely, i) nodes: to represent each single feature in a domain, ii) edges: to connect nodes by directed links, which determine the conditional dependence relationship between them, and iii) conditional probability tables (CPTs): to list the probability distribution of each node (represented feature) in relation to the other connected nodes [14].

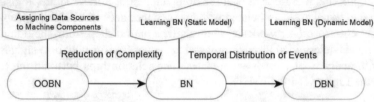

**Fig. 4.** 3-Step Procedure for Creating and Learning DBN

A DBN, as a special case of BN, changes the nature of a static BN to a dynamic model through the evolution of random variables and temporal distribution in discrete time points $i$ in the interval of $0 \le i \le T$. Stationary DBNs, in contrast to the non-stationery ones e.g. discussed in [15],[16], do not consider evolving network structure (changes of edges) over time. This is a valid hypothesis in the maintenance of CPPS in which the cause-effect relationships remain stationary, excluding highly flexible production system and full automation (i.e. the highest autonomy level of CPPS). By integrating the Event-Cost Schema with DBNs we aim at unifying temporal dimension of cause-effect analysis with uncertainty and estimating the potential implications on production planning and maintenance costs. To that end, the 3-step procedure for creating and learning DBN presented in [6] is employed as follows: i) assigning variables (e.g. failure rate) to real objects (e.g. machine components) using Object-Oriented Bayesian Networks (OOBN), ii) creating or learning a BNs (static model) based on a given number of observations, and iii) converting BNs to DBNs (dynamic model) by adding the temporal dimension of events (Cf. Figure 4). Manual construction of DBNs requires domain expert knowledge to build the networks (nodes and edges) and then assess the conditional probability distributions. Automatic learning of DBNs from the Event-Cost Schema, however, requires using certain algorithms such as the Expectation Maximization (EM) or General Expectation Maximization (GEM) [17]. Creating or learning a DBN enables us not only to compute (inference) values of each node $x_i$ in every time slice in relation to the other conditioned nodes (connected via edges), but also to predict the future states (values) of nodes $x_{t+1}$ based on the past observations. Given the values of the past observation nodes in every time slice $Y_0^t$, we should estimate the hidden values of nodes $x_i$ with the maximum a posterior probability (maximum likelihood (ML)) using the algorithm 1 (Cf. Figure 5).

---

```
Input: Past Observation values
(1) Past observation time slices Y₀ᵗ ∈ {y₀  y₁ ... yₜ}
(2) Past observation values of nodes xᵢ ∈ {x₀  x₁ ... xₜ}
Output: Prediction of values of nodes xₜ₊₁
for each time t+1 collect the past observation data do
```
$$x_{t+1,tML} = \arg\max_{x_{t+1}} \Pr(x_{t+1}|Y_0^t) \quad \text{and} \quad y_{t+1,tML} = \arg\max_{y_{t+1}} \Pr(y_{t+1}|Y_0^t)$$
```
Update the observation values xᵢ and Y₀ᵗ in DBNs.
```

---

**Fig. 5.** Algorithm 1 for using DBNs to Make Prediction – Adopted from [16] and [17]

## 4    Conclusion and Future Research Agenda

The undertaking paradigm shift towards smart production systems lead to evolution of maintenance strategies from diagnostic to prescriptive. This paper proposes the PriMa model as an advancement of the predictive model presented in the context of research project Maintenance 4.0. On the conceptual level, we have elaborated on creating and learning dynamic models for predicting future events and prescribing optimal action plans considering multimodality and structural heterogeneity of maintenance data. Furthermore, we have discussed the methodology for creating and learning DBNs using multimodal maintenance records.

The future research agenda is firstly to realize the integrated PriMa model and to evaluate its effectiveness using shop floor and production planning data in cooperation with industry partners. Secondly, we examine the integration of the proposed model in the existing production planning systems such as MES and ERP. Thirdly, to go beyond the stationary DBNs, we should investigate learning non-stationary DBNs for maintenance of highly flexible production systems, e.g. in semi-conductor manufacturing systems. Finally yet importantly, the underlying data model can be transformed to a semantic knowledge-base for continuous learning and improving (i.e. supporting reasoning and problem-solving tasks) in maintenance planning, monitoring and controlling.

## Acknowledgement

The authors would like to acknowledge the financial support of the Austrian Research Promotion Agency (FFG) for funding the research project of "Maintenance 4.0" (2014-2017) under the grant number 843668.

## References

1. Azadeh, A., Salehi, V., Jokar, M., and Asgari, A., 2016. An Integrated Multi-Criteria Computer Simulation-AHP-TOPSIS Approach for Optimum Maintenance Planning by Incorporating Operator Error and Learning Effects. *Intelligent Industrial Systems*, 2 (1), pp. 35-53.

2. Windmann, S., Eickmeyer, J., Jungbluth, F., Badinger, J., and Niggemann, O., 2016. Evaluation of Model-Based Condition Monitoring Systems in Industrial Application Cases. In Machine Learning for Cyber Physical Systems. Springer, Berlin Heidelberg, pp. 45-50.

3. Li, P., and Niggemann, O., 2016. Improving Clustering based Anomaly Detection with Concave Hull: An Application in Fault Diagnosis of Wind Turbines. In 2016 IEEE 14th International Conference on Industrial Informatics (INDIN), IEEE Press, pp. 463-466.

4. Paschke, F., Bayer, C., and Enge-Rosenblatt, O., 2016. A Generic Approach for Detection of Wear-out Situations in Machine Subsystems. In 2016 IEEE 21st International Conference on Emerging Technologies and Factory Automation (ETFA), IEEE Press, pp. 1-4.

5. Ansari, F., Uhr, P., and Fathi, M., 2014. Textual Meta-analysis of Maintenance Management's Knowledge Assets. *International Journal of Services, Economics and Management*, 6 (1), pp. 14-37.

6. Dienst, S., Ansari, F., and Fathi, M., 2015. Integrated System for Analyzing Maintenance Records in Product Improvement. *The International Journal of Advanced Manufacturing Technology*, 76 (1-4), Springer, pp. 545-564.

7. Ansari, F., and Fathi, M., 2016. Meta-analysis of Maintenance Knowledge Assets towards Predictive Cost Controlling of Cyber Physical Production Systems. In Machine Learning for Cyber Physical Systems. Springer, Berlin Heidelberg, pp. 103-110.

8. Glawar, R., Kemeny, Z., Nemeth, T., Matyas, K., Monostori, L., and Sihn, W., 2016. A Holistic Approach for Quality Oriented Maintenance Planning Supported by Data Mining Methods. Procedia CIRP, Vol. 57, pp. 259 - 264.

9. Flatt, H., Koch, N., Röcker, C., Günter, A., & Jasperneite, J., 2015. A Context-aware Assistance System for Maintenance Applications in Smart Factories based on Augmented Reality and Indoor Localization. In 2015 IEEE 20th Conference on Emerging Technologies & Factory Automation (ETFA), IEEE Press, pp. 1-4.

10. Wang, J., Zhang, L., Duan, L., and Gao, R. X., 2017. A New Paradigm of Cloud-based Predictive Maintenance for Intelligent Manufacturing. *Journal of Intelligent Manufacturing*, 28 (5), pp. 1125-1137.

11. Abramovici, M., Gebus, P., Göbel, J. C., and Savarino, P., 2017. Provider-Driven Knowledge Allocation Concept for Improving Technical Repair Tasks in IPS 2 Networks. Procedia CIRP, Vol. 64, pp. 381-386.

12. Bärenfänger-Wojciechowski, S., Austerjost, M., and Henke, M., 2017. Smart Maintenance-Asset Management der Zukunft: Ein integrativer Management-Ansatz (Smart Maintenance - Asset Management for the Future: An Integrative Management Approach). wt-online 1/2-2017, Springer-VDI Verlag, pp. 102-106.

13. Linstedt, D., and Olschimke, M., 2015. Building a Scalable Data Warehouse with Data Vault 2.0, 1st Edition, Elsevier, pp. 17-32 and 89-121.

14. Kelleher, J. D., Mac Namee, B., and D'Arcy, A., 2015. Fundamentals of Machine Learning for Predictive Data Analytics: Algorithms, Worked Examples, and Case Studies, MIT Press, pp. 292-306.

15. Gonzales, C., Dubuisson, S., and Manfredotti, C. E., 2015. A New Algorithm for Learning Non-Stationary Dynamic Bayesian Networks with Application to Event Detection. In FLAIRS Conference, pp. 564-569.

16. Robinson, J. W., and Hartemink, A. J. 2010. Learning Non-Stationary Dynamic Bayesian Networks. *Journal of Machine Learning Research*, 11, pp. 3647-3680.

17. Mihajlovic, V., and Petkovic, M., 2001. Dynamic Bayesian Networks: A State of the Art. Technical Report, University of Twente, The Netherlands.

# Evaluation of Deep Autoencoders for Prediction of Adjustment Points in the Mass Production of Sensors

Martin Lachmann[1], Tilman Stark[1], Martin Golz[2], Eberhard Manske[3]

[1] Robert Bosch Fahrzeugelektrik Eisenach GmbH, Robert-Bosch-Allee 1, 99817 Eisenach
martin.lachmann@de.bosch.com
[2] Hochschule Schmalkalden, Fakultät Informatik, Blechhammer 4-9, 98574 Schmalkalden
[3] Technische Universität Ilmenau, Fakultät Maschinenbau, PF100565, 98684 Ilmenau

**Abstract.** In the context of Industry 4.0 the inclusion of additional information from the manufacturing process is a challenging approach. This is demonstrated by an example of calibration process optimization in the mass production of automotive sensor modules. It is investigated to replace a part of a measurement set by prediction. Support-vector regression compared to multiple, linear regression model shows only minor improvements. Feature reduction by deep autoencoders was carried out, but failed to achieve further improvements.

**Keywords:** machine learning, industry 4.0, deep autoencoder, support vector regression, feature reduction, alignment process, production data

## 1  Introduction

For more than ten years, deep learning networks are evolving as an essential part of neuroinformatics. Important paradigms are deep autoencoders (DAE), deep belief networks and convolutional networks. Major breakthroughs have been achieved in the fields of object, scene and speech recognition as well as automatic translations and genomics. An interesting application of DAE for feature reduction with support vector machines (SVM) as classification method has been presented [1] resulting in significantly improved accuracies. This contribution presents a similar concept to a data set from the mass production of automotive sensor modules. Instead of SVM, support vector regression (SVR) is used for regression analysis.

The aim of these investigations is the prediction of cost intensive measurements using previously recorded process and quality data. These can be used for individual characteristics and temperature behaviors compensation. Several simplifications of this process, e.g. calibration of pressure sensors, have been evaluated [2]. However, the described approaches of changing the alignment model and removing measurements without replacement did not achieved the desired results of accuracy and stability of calibration. Here it is aimed to reduce measurement complexity by introducing an appropriate and reliable predictive model without a negative effect on the process results. Difficulties were expected due to measurement uncertainty and their consistency across a wide range of production stations. The production of automotive sensor modules places high requirements to the maximum errors and to the stability

© Springer-Verlag GmbH Germany, part of Springer Nature 2020
J. Beyerer et al. (Eds.), *Machine Learning for Cyber Physical Systems*, Technologien für die intelligente Automation 11,
https://doi.org/10.1007/978-3-662-59084-3_2

due to tolerable error rates, which are typically below 1 ppm. An incorrectly predicted measure value has direct impact to the sensor error and may finally violate the product tolerances.

Since individual modules are not sufficiently described by the model, the focus was on maximum errors instead of mean errors. There is hope that the inclusion of less relevant variables can lead to an improvement.

## 2    Material and Methods

### 2.1    Use Case

The dataset includes measurements of 37.500 automotive sensors modules of a single type. Measurements originate from a calibration process of two measured variables: temperature $T$ and another one, afterwards referred to as $X$ for reasons of confidentiality. Both variables are recorded at 3 alignment points $S_{00}$, $S_{01}$ und $S_{10}$ (Fig. 1) to find parameters of a polynomial alignment model processed within the digital signal processor of the module (Fig. 2). The aim of our data analysis is to predict the measured variable $X$ at alignment point $S_{11}$. The motivation for this is an increased cost efficiency of the calibration process.

In addition to the 6 measurements $T(S_{00}), T(S_{01}), T(S_{10}), X(S_{00}), X(S_{01}), X(S_{10})$ several variables of the production process of the automotive sensor modules were recorded and 169 features were extracted. Some of the features refer to temporal and spatial locations used for traceability at different stations, especially in single stations within multiple production or test locations.

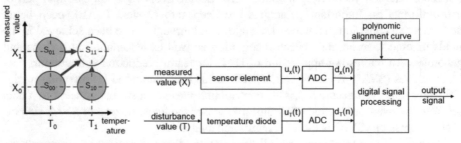

**Fig. 1** (left): Alignment points of measurements (blue) and prediction (red)
**Fig. 2** (right): Block diagram of signal processing

Two regression models were considered in this contribution:
a) Multiple, linear model

$$X(S_{11}) = p_0 + p_1 T(S_{00}) + p_2 T(S_{01}) + p_3 T(S_{10}) + p_4 X(S_{00}) + p_5 X(S_{01}) + p_6 X(S_{10}) \quad (1)$$

The parameter vector $p = (p_0, p_1, ..., p_6)^T$ has to be estimated empirically. As a result, the model is able to predict the temperature dependency of measured value $X$ at different operation points within the interval $[S_{00}, S_{10}]$. To some extent, the model will also be able to extrapolate outside the interval.

b) Multivariate, linear and non-linear regression

SVR (ch. 2.2) is used to predict $X(S_{11})$. The input variables were the 6 variables of the linear model and up to 169 additional features mentioned above. If SVR with no kernel is used, a linear model is adapted to the distribution of the input variables. The usage of kernels leads to an implicitly expressed transformation from the input space to the reproducing kernel Hilbert space (RKHS), where then again a linear model has to be adapted. Within the input space, this represents a non-linear regression.

Both measured variables $X(S_{01})$ und $X(S_{11})$ were almost normally distributed with a few outliers. To increase the sensitivity of the regression analysis for problematical samples at the margin of the distribution, a subsample was extracted, consisting of 67 % of elements of the marginal region and 33 % of the central region of the distribution. Values lower than the 5th percentile or higher than the 95th percentile of both univariate distributions were considered as marginal regions. For both distributions, the intersection was about 17.8 % of all 37,500 sampling elements. About 8.9 % were drawn from the central region and were included resulting in a relation between marginal and central elements of 2:1. This way, the subsample puts more emphasis to the optimization of the maximum error than to the mean square error.

## 2.2    Deep Autoencoder

In a deep neural network, the neurons of the first layer produce non-linear representations of the input vector (see Figure 3). In the second layer, neurons produce non-linear representations of the representations of the first layer and thus represent a higher level of abstraction; and so on for the next layers. In the DAE network, two aspects are characteristic. Firstly, there is a layer, the so-called code layer, which has significantly fewer neurons than the previous layers. Secondly, the architecture is mirror symmetric to the central code layer. Beyond this layer are layers with the same number of neurons and transposed weight matrices. Thus, the last layer has the same number of neurons as the input layer, which is equal to the dimensionality of the input vector. DAE training aims at generating the output vector $\hat{x}_i$ as an approximation of the input vector $x_i$. During the training process, the mean squared error (MSE) between $x_i$ and $\hat{x}_i$ is optimized. At the end of the training process, the code layer generates low-dimensional vectors $g_i$ of the input vectors $x_i$; $g_i$ represents values required to reconstruct $x_i$. Thus, a non-linear dimension reduction is achieved well depending on MSE.

Training deep neural networks poses high challenges both to algorithms as well as to the computing load. It was carried out layer-wise with "restricted Boltzmann machine" (RBM) and "enhanced gradient" to achieve high convergence rate and a more reliable MSE minimum [3]. Firstly, the weight matrices $W_{D1}$, $W_{E1}$ and the threshold values of the first layer neurons were optimised by RBM, under the restriction that $W_{E1} = W_{D1}^T$ holds [4]. Afterwards, $W_{D2}, W_{E2}$ under restriction $W_{E2} = W_{D2}^T$ were optimised, and so on. Finally, the whole DAE net was tuned within the error backpropagation paradigm, whereby restrictions were abandoned: $W_{Ei} \neq W_{Di}^{\mathrm{T}} \forall\, i = 1, ..., N_S$ with $N_S$ as number of layers up to the code layer ($N_S = 3$, Figure 3)

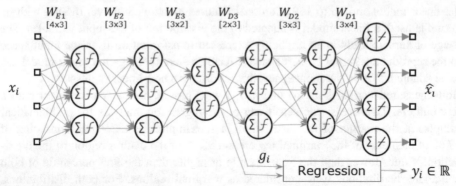

**Fig. 3.** Simple DAE network with 4-3-3-2 architecture to compress high-dimensional feature vectors $x_i$ to low-dim. representation vectors $g_i$ which are mapped to the target $y_i$.

## 2.3    Support vector regression

Based on the ideas of statistical learning theory [5], support vector regression (SVR) [6] is a method of multivariate regression generating a mapping $f: x \mapsto y$ from feature vectors $x \in \mathbb{R}^d$ to a target variable $y \in \mathbb{R}$ utilizing machine learning. This requires a training     set     $T$     with     sample     size     $n$     as     large     as     possible.

$$T = \{(x_i, y_i) | x_i \in \mathbb{R}^d, y_i \in \mathbb{R}, i = 1, ..., n\}. \tag{1}$$

It is aimed at finding all parameters of $f(x)$ which has absolute deviations of $f(x_i)$ and $y_i$ less than an upper error bound $\varepsilon$ for all pairs $(x_i, y_i) \in T$. Deviations smaller than $\varepsilon$ will be accepted without contributions to the loss function to be minimized. Deviations larger than $\varepsilon$ lead to linearly increasing contributions to the loss function. This concept is called $\varepsilon$-insensitive loss [6]. For a multivariate regression with the linear function

$$f(x) = \langle w, x \rangle + b\,, b \in \mathbb{R}\,, w \in \mathbb{R}^d \tag{2}$$

it has been shown [6], that the norm $\|w\|$ must be minimized under constraints due to the $\varepsilon$-insensitive loss. This optimization problem holds if a monotonous function of $\|w\|$ is minimized instead of $\|w\|$:

$$\begin{aligned} &\text{minimize w.r.t. } w, b: \quad \tfrac{1}{2}\|w\|^2 \\ &\text{subject to:} \quad |y_i - \langle w, x_i \rangle - b| \leq |\varepsilon| \end{aligned} \tag{3}$$

This convex optimization problem must be admissible, i.e. for all pairs $(x_i, y_i) \in T$ equ. (3) holds. Otherwise, slack variables $\xi_i, \xi_i^*$ have to be introduced, which allow violations of (3) in positive and negative direction, respectively, for some examples $i \in \{1, ..., n\}$. Solutions of (3) can be found by Lagrangian optimization. Once Lagrange multipliers $\alpha_i, \alpha_i^*$ were found, the solution of $w$ with $x_i \in T$ follows:

$$w = \sum_{i=1}^{n}(\alpha_i - \alpha_i^*)x_i \tag{4}$$

This is a sparse solution, because for all feature vectors inside the $\varepsilon$-tube (3) their multipliers $\alpha_i, \alpha_i^*$ disappear. Only feature vectors $x_i$ with $\alpha_i \neq 0, \alpha_i^* \neq 0$ contribute to the solution (4). They are called support vectors. By substituting (4) in (2) yields:

$$f(x) = \sum_{i=1}^{n}(\alpha_i - \alpha_i^*)\langle x_i, x \rangle + b \tag{5}$$

To have a nonlinear model, a nonlinear transformation to the reproducing kernel Hilbert space (RKHS) utilizing kernel functions has been introduced [7]. The inner product $\langle x_i, x \rangle$ is now replaced by an admissable kernel $k(x_i, x)$. Such kernels represent an inner product of nonlinear function vectors $k(x_i, x) = \langle \varphi(x_i), \varphi(x) \rangle$ which contains all eigenfunctions of $k$ and are elements of the RKHS. The dimensionality of the eigenfunction vectors $\varphi$ is much higher than the dimensionality $d$ of the feature space. For example, if a polynomial kernel is used, the RKHS dimensionality is equal to the number of monomials after which the polynomial of order $p$ can be expanded and is $\binom{d+p-1}{p}$. Given a polynomial order of e.g. $p = 6$, feature vectors of dimensionality $d = 10$ are represented in the RKHS of dimensionality 5005. This is of great advantage since in high-dimensional spaces the number of degrees of freedom is correspondingly high and as well the chance to find a linear solution in RKHS. An explicit solution of $w$ like in (4) is not feasible to formulate, but for $f(x)$ like in (5):

$$f(x) = \sum_{i=1}^{n}(\alpha_i - \alpha_i^*)k(x_i, x) + b \tag{6}$$

Examples of admissible kernels can be found in [8], as well as their necessary and sufficient conditions. In this contribution, the translation-invariant radial basis function kernel (RBF) with kernel width $\sigma$ as a free parameter is used:

$$k(x, x_i) = \exp\left(-\frac{\|x - x_i\|^2}{2\sigma^2}\right). \tag{7}$$

## 3    Results

This chapter summarizes the results for all three investigated data sets. The first subset contains both measured quantities $X$, $T$ at 3 different adjustment points. The second subset contains the first and 34 additional and highly relevant features of data from the production line, which were subjectively selected by a domain expert. The third subset consists of the second and all features not selected by the expert. The mean squared error (MSE) between the predicted and the measured $X(S_{11})$ has been used as validation criterion. Additionally, the difference (full-scale error) between measured and predicted sensor output value is used. The full-scale error are calculated by doing a virtual characterization, using a model of a sensor signal path. In our use case the maximum MSE is 8.22, it is given by predicting $X(S_{11})$ using a polynomial model $0^{th}$ order. 5-fold cross validation was used as validation method. Best results for SVR has been achieved after numerically extensive hyperparameter optimization, realized by EPSGO [9].

With the first subset, the three methods were compared (Table 1). With only 6 feature a further reduction does not make sense and is therefore omitted.

| Regression method | Training MSE | Test MSE | Mean full-scale error | Maximum full-scale error |
|---|---|---|---|---|
| Linear model | 2,38 | 2,55 | 0,11% | 0,43% |
| SVR (linear kernel) | 2,46 | 2,50 | 0,11% | 4,31% |
| SVR (RBF kernel) | 2,50 | 2,50 | 0,11% | 4,36% |

Table 1. Comparison of regression methods based on data subset 1.

In the second case, 40 features selected by domain experts are used. This contains 6 features from the calibration process, numbers of test positions and features from build-in self-test. For every test position, one binary valued feature was applied. The SVR with RBF kernel was used as regression method (Table 2).

| Feature reduction | $n_f$ | Training MSE | Test MSE | Mean error full scale | Max. full scale error |
|---|---|---|---|---|---|
| no reduction | 40 | 2,46 | 2,57 | 0,11% | 5,31% |
| deep autoencoder (linear) | 20 | 8,05 | 8,06 | 0,12% | 7,73% |
| deep autoencoder (tansig) | 20 | 8,05 | 8,06 | 0,13% | 7,78% |
| PCA | 20 | 7,45 | 7,65 | 0,14% | 7,65% |

Table 2. Results of regression using 40 features selected by domain experts ($n_f$ is the number of features).

The last case, i.e. subset 3, contains all 175 features available from subset2 as well as testing and calibration. The SVR with RBF kernel was used as regression method (Table 3).

| Feature reduction | $n_f$ | Training MSE | Test MSE | Mean error full scale | Max. full scale error |
|---|---|---|---|---|---|
| no reduction | 175 | 1,13 | 3,56 | 0,11% | 4,39% |
| deep autoencoder (linear) | 80 | 4,86 | 5,90 | 0,15% | 7,79% |
| deep autoencoder (tansig) | 80 | 4,98 | 6,12 | 0,15% | 7,79% |
| PCA | 80 | 5,71 | 7,38 | 0,14% | 8,19% |

Table 3. Results with all 175 parameters ($n_f$ – number of features for regression)

## 4      Discussion

It has been shown that prediction quality by comparing MSE is slightly improved using SVR instead of the former existing linear model when the first data set consisting only of the six measured values, i.e. $T(S_{00}), T(S_{01}), T(S_{10}), X(S_{00}), X(S_{01}), X(S_{10})$ was used.

Inclusion of the full set of 175 features did not lower the error rates. The same applies to the second data set of 40 features chosen by a domain expert. It appears that the additional features introduce a high amount of non-relevant information to the analysis, causing an increase in prediction error rates. Despite this fact, these features were left in the analysis to achieve improved robustness to outlier.

With this intention, PCA and DAE were used for automated feature reduction. PCA is based on a linear model for the Gaussian distribution, whereas DAE is nonlinear and is not restricted to a particular distribution. Results show drastically increased prediction errors with both feature reduction. An investigation of this unexpected result showed major differences in feature relevancies, especially all 6 measured variables. Two of them are highly relevant. The error sensitivity due to additional white noise of these highly relevant features (Fig. 4) was correspondingly high, such that even small reconstruction errors of these variables during DAE and PCA led to major regression error rates. Furthermore, it turned out that relatively small deviations in individual outlier values had a degrading influence. The outliers led to a concentration of most of the samples within a small interval (Fig. 5).

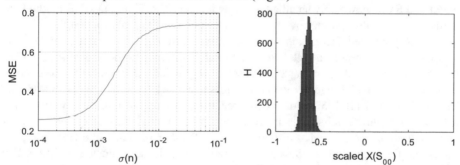

**Fig. 4 (left)**: Noise sensitivity of $X(S_{00})$: MSE of regression vs. standard deviation of added white noise

**Fig. 5 (right)**: Empirical distribution of scaled variable $X(S_{00})$.

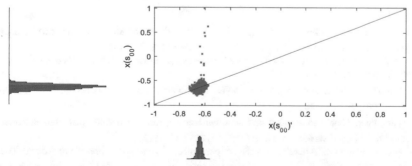

**Fig. 6**: Scatter diagram and empirical distributions of $X(S_{00})$ and the reconstruction $\hat{X}(S_{00})$

DAE aims at minimizing the mean reconstruction error, which is most influenced by variables with large variance. For example $X(S_{00})$ has relatively low variance (Fig. 5) and therefore affects the MSE relatively small. The correlation between $X(S_{00})$ and his reconstruction $X(S_{00})$' is bad (Fig. 6).

An improvement of the reconstruction has been achieved by scaling. DAE with linear activation functions can create values outside an interval (-1, 1), whereas DAE with tansig activation function cannot. This has already made a significant improvement, but did not exceed SVR without feature reduction.

The evaluation of maximum full-scale errors indicates, that neither the linear model nor the SVR is accurate enough to meet the tolerable error of less than 0.3 %FS. Nevertheless, the comparison of both model enables a new approach, because every outliers in the subset was predicted well enough by one of the two model. Large differences in the predicted values of both models enables a validation and can also be applied for anomaly detection.

# 5     Conclusions

A first evaluation of predictions made for a calibration procedure on data of a production line of automotive sensor modules. The comparison between a multiple, linear model and SVR shows only small advantages for the SVR. Using DAE as method for feature reduction did not improve results. Subsequent analysis shows that the relevancies of the processed features is very unevenly distributed. The training of DAE is unsupervised and represent all features. Small additional random errors in highly relevant features resulted in major increases of regression error rates. One solution is to use the linear model together with SVR. It is necessary to go in detailed analysis of the distribution of all 175 features extracted from production line data. Further improvements could be achievable by recording of additional data of production process. Because of the presented difficulties of SVR other methods like Gaussian processes and the suggested combination two models should be evaluated in future. For subsequent evaluations an extended sample quantity should be considered to characterize outliers much better and to prove robustness of the final model for millions of annual produced parts.

# References

[1] Vincent, Larochelle, Lajoie, Bengio, Manzagol (2010) Stacked denoising autoencoders: Learning useful representations in a deep network. *J Machine Learning Research* 11: 3371-3408

[2] Možek, et al. "Adaptive Calibration and Quality Control of Smart Sensors." *Applications and Experiences of Quality Control*. InTech, 2011.

[3] Bengio, Lamblin, Popovici, Larochelle. Greedy layer-wise training of deep networks. Adv neural inform proc syst (NIPS) 2007;19:153–160.

[4] Cho, Tapani, Ihler. Enhanced gradient and adaptive learning rate for training restricted Boltzmann machines. Proc 28th Int Conf Machine Learning (ICML-11) 2011:105–112.

[5] Vapnik, and Chervonenkis. "A note on one class of perceptrons." *Automation and remote control* 25.1 (1964): 103.

[6] Vapnik. "The Nature of Statistical Learning Theory." (1995).

[7] Aronszajn. "Theory of reproducing kernels." Trans American mathem society 68.3 (1950): 337-404.

[8] Smola , et. al. "A tutorial on support vector regression." Statistics and computing 14.3 (2004): 199-222.

[9] Fröhlich, Zell. "Efficient parameter selection for support vector machines in classification and regression via model-based global optimization." *Neural Networks, 2005. IJCNN'05. Proceedings. 2005 IEEE International Joint Conference on*. Vol. 3. IEEE, 2005.

# Differential Evolution in Production Process Optimization of Cyber Physical Systems

Katharina Giese[1], Jens Eickmeyer[1], and Oliver Niggemann[1,2]

[1] Fraunhofer IOSB-INA, Lemgo
[2] Institute Industrial IT (inIT), Lemgo

**Abstract.** In this paper, the application of Differential Evolution in machine optimization is introduced. This enables the optimization of different production processes in modern industrial machines, without having in depth knowledge of the inner workings of production units. Therefor, sensor data is recorded and certain properties like manufacturing time or quality are introduced as new fitness criteria for the evolutionary computing algorithm. This is demonstrated in an exemplary use case for injection moulding. Furthermore, a concept for constant production process stabilization is presented for future research.

## 1 Introduction

Following modern production standards, Industry 4.0 machines enable users to customize production processes according to their specific needs. Corresponding industrial machines allow to configure settings like temperature, pressure or operating speed while recording the actual manufacturing process through sensors for later diagnostics.

Due to the vast amount of different possible configurations, operators are facing the challenge of determining and adjusting ideal settings. As a consequence, they are required to have an extensive knowledge of the production process as well as the machine's performance and characteristics.

This poses a problem, as managing the production plant becomes more difficult with the size, demanding time and resources. Thereby, the solution lies within the automation of setting configuration through constant condition monitoring and machine optimization.

This work presents an implementation of the Differential Evolution (DE) algorithm to analyze and model sensor data collected from cyber physical systems, opening up new opportunities to identify and continuously optimize a production unit's performance, as depicted in Figure 1.

A DE implementation not only enables performance optimization regarding criteria like production time, power consumption or quality, but also offers the possibility to stabilize this process by subtracting age-related impacts through runtime observation. Furthermore, the algorithm is universally applicable to various kinds of production machinery without the need to have an in depth knowledge of the inner workings.

© Springer-Verlag GmbH Germany, part of Springer Nature 2020
J. Beyerer et al. (Eds.), *Machine Learning for Cyber Physical Systems*, Technologien für die intelligente Automation 11,
https://doi.org/10.1007/978-3-662-59084-3_3

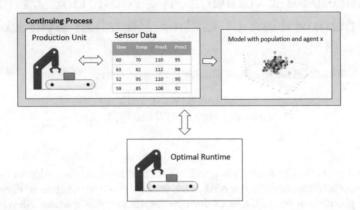

**Figure 1.** Application of Differential Evolution

This paper is structured as follows: Section 2 outlines the research field and contribution of this paper. DE is described and modified for machine optimization in Section 3. Section 4 provides an evaluation of an exemplary use case extending to a concept of runtime optimization in Section 5. Finally, Section 6 gives a conclusion and states further research possibilities.

## 2    Related Work and Contribution

DE can be classified as an evolutionary algorithm, which was first introduced in an article by Storn and Price in 1995 [1]. It is primarily used for numerical problem optimization over a continuous search space while demonstrating a relatively short runtime. This has been proven at the First International Contest on Evolutionary Computation, where DE came in third, while being the first universally applicable algorithm to finish the task [2]. It also shows a great performance compared to particle swarm optimization and evolutionary algorithms as indicated in a study by Vesterstrom and Thomsen [3].

In recent years, different DE variants have been introduced and applied for diverse use cases, which include the training of neural networks, power system planning and pattern recognition [4]. Furthermore, methods for additional runtime optimization have been developed, such as extending DE by the prediction of nearest neighbours through training a kNN predictor by Liu and Sun [5].

Also, industrial use-cases have been solved with the help of evolutionary algorithms [6], thus laying the foundation for enhancing optimization processes through the use of DE.

In this research, DE's usage in industrial production plants is examined, leading to the following main contributions:

– Modifying the algorithm for optimal parameter estimation of industrial machinery

– Verifying the functionality through an exemplary use case
– Devising a concept for continuous optimization through condition monitoring

## 3    Process Optimization with Differential Evolution

In this Section, the operating principle of DE is explained, as well as its application in machine optimization.

### 3.1    Operating Principle of Differential Evolution

As an evolutionary algorithm, DE is based on real biological evolutionary processes. Those are marked by four characteristics:

– Initialization of a starting population: Converting data from production cycles to population members which are used to start the optimization iteration
– Inheritance from parent population members leading to mutations: Randomly chosen production cycles provide the foundation for recombination
– Recombination of certain factors with a candidate solution: Sensor data from the chosen cycles is combined to a new artificial production cycle via binary crossover
– Selection via a fitness function: Assessment of the newly created production cycle in relation to the existing population

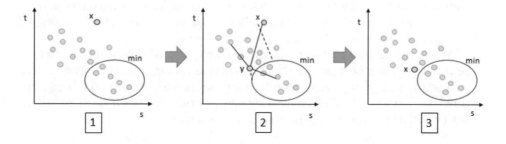

**Figure 2.** Operating Principle of DE

The starting population is initialized as a list of agents representing vectors with certain components as shown in Figure 2.1. According to the number of different components, those agents can be visualized building the population in a n-dimensional space. The first candidate solution, posing as the individual of the population which is to be evolved, is generated using random numbers within the boundaries of the original components.

By using randomly chosen agents from the existing population and the first candidate solution x, a second candidate solution y is created through binary crossover, representing the exchange of genetic material of chromosomes in real biological evolutionary processes. Thus, the second candidate solution inherits attributes from each of their "parents", taking into account adjustable crossover possibilities, see Figure 2.2.

In the final step, both candidate solutions are compared according to a fitness function determining the evolutionary fitter candidate. This candidate will be used as the new candidate solution x for further recombination processes, as depicted in Figure 2.3.

## 3.2   Modification for Machine Optimization

When applied to machine optimization, the population is build using a machine's recorded sensor data. Meaning, in this case the population members are vectors derived from each production cycle, whereas the sensor data acts as the vector components. Through the iterative combination of those vectors with the optimal candidate solution, the ideal configuration for a production process can be achieved.

Because in machine optimization, there is no numerical fitness function per se, certain criteria have to be chosen to classify the evolutionary fitness of vectors. Those criteria must be measurable, for instance power or material consumption, reject rate or the quality rated on a numerical scale. In this case, the aim is to reach a stable production through the optimization of the production cycle time.

The candidate solution doesn't have the classifying attribute, since its configurations are not yet tested in the machine. Therefore, for optimization, the candidate solution with the fittest vector in its immediate environment is used. If its the same with x and y, the smallest distance counts. This is measured through the x-dimensional euclid or using a kd-tree.

After a set number of iterations, the momentary candidate solution works as the result, showing best parameter configuration through its components.

When wanting to optimize several properties at once, a method of weighting should be introduced. This enables setting priorities in production processes, thus providing the option for an overall optimization.

## 4   Evaluation

To prove its functionality via reverse engineering, the modified DE algorithm has been implemented in Python 3.6. For this reason, a strategically chosen artificial test data set has been generated based on a predefined optimal configuration. The aim is to retrieve this optimum by running DE.

The prepared test data set meets the following conditions:

- The parameters and values are based on real life injection moulding data sets
- Correlations exist within parameters like temperature, cooling time, power consumption etc.

- Noise and random leaps have been added to improve realism
- The optimal configuration vector is not available in the test data set

The following vector is the optimal configuration vector, on which the test data set is build on, showing the settings necessary to reach the minimum cycle time of 10 seconds:

| Temp1 | Temp2 | Press1 | Press2 | Cool | Power | Dense |
|-------|-------|--------|--------|------|-------|-------|
| 185   | 43    | 990    | 900    | 1.2  | 0.55  | 1.2   |
| Pos1u | Pos1d | Pos2u  | Pos2d  | Pos3u | Pos3d |      |
| 10    | 6     | 4      | 1.5    | 5.7  | 3.11  |       |

Its first seven values are correlating while the last six are chosen randomly to demonstrate the algorithms ability to neglect irrelevant data.

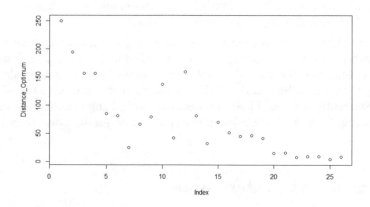

**Figure 3.** Convergence of Candidate Solution to the Optimum Vector

Figure 3 shows the distance of the current candidate solution to the defined optimal configuration vector in an euclidean space, only considering the first seven values, since the random ones are irrelevant to this evaluation. As seen in the scatter plot, it starts approaching the optimum quickly, fluctuates around seven to twelve iterations and then levels of around a distance of 8.33. The fluctuation can be explained as local optimums, which the algorithm approximates while searching for the global optimum.

The following table shows the relevant data of four runs in comparison to the optimal configuration vector:

| | Temp1 | Temp2 | Press1 | Press2 | Cool | Power | Dense |
|-----|-------|-------|--------|--------|------|-------|-------|
| Opt | 185 | 43 | 990 | 900 | 1.2 | 0.55 | 1.2 |
| 1 | 181 | 44 | 925 | 847 | 1.4 | 0.35 | 1.1 |
| 2 | 184 | 42 | 975 | 912 | 1.4 | 0.44 | 1.1 |
| 3 | 191 | 42 | 991 | 898 | 1.3 | 0.51 | 1.2 |
| 4 | 186 | 43 | 1001 | 892 | 1.1 | 0.62 | 1.4 |

This evaluation proves the expected functionality of the proposed modifications for machine optimization, as the output is similar to the optimum vector.

## 5  Continuous Optimization

For continuous optimization, this setup has to be expanded using a time parameter. The data set will then be updated regularly and older points will be weighted more severely. Therefore, the older the received stats are at a certain point in time, the lower will be their fitness, counteracting malfunctions of the machine, which may develop in the course of time.

As the implemented algorithm is only aware of the stats of the closest population members and the weighting of those found unfit is executed following an exponential function, the candidate solution is more likely to follow along the newer population members and stabilizing the current state of the production machine concerning wear and tear. This enables optimizing machines on runtime without having in depth knowledge of the machine or production process.

## 6  Conclusion and Future Work

To conclude, DE provides a fast and resource efficient way to optimize production processes following certain criteria. The traditional fitness function is replaced by analyzing nearby population members, thus facilitating the use in machine optimization.

The application on artificial injection moulding data proved to be successful demonstrating a close to correct parameter estimation. This paves the way for further research on continuous optimization and stabilization procedures at runtime, made possible by incorporating timestamps and penalty for elderly production cycles.

The aim is to ultimately provide a quick and easy way to stabilize production processes of unknown industrial machines.

# References

1. R. Storn and K. Price, "Differential Evolution - A simple and efficient adaptive scheme for global optimization over continuous spaces," *Journal of Global Optimization*, pp. 1–15, 1995.
2. H. Bersini, M. Dorigo, S. Langerman, G. Seront, and L. Gambardella, "Results of the first international contest on evolutionary optimisation (1st iceo)," in *Evolutionary Computation, 1996., Proceedings of IEEE International Conference on.* IEEE, 1996, pp. 611–615.
3. J. Vesterstrom and R. Thomsen, "A comparative study of differential evolution, particle swarm optimization, and evolutionary algorithms on numerical benchmark problems," in *Proceedings of the 2004 Congress on Evolutionary Computation (IEEE Cat. No.04TH8753)*, vol. 2, June 2004, pp. 1980–1987 Vol.2.
4. S. Das and P. N. Suganthan, "Differential evolution: A survey of the state-of-the-art," *IEEE Transactions on Evolutionary Computation*, vol. 15, no. 1, pp. 4–31, Feb 2011.
5. Y. Liu and F. Sun, "A fast differential evolution algorithm using k -Nearest Neighbour predictor," *Expert Systems With Applications*, vol. 38, no. 4, pp. 4254–4258, 2011.
6. A. Diedrich, J. Eickmeyer, P. Li, T. Hoppe, M. Fuchs, and O. Niggemann, "Universal Process Optimization Assistant for Medium-sized Manufacturing Enterprises as Self-learning Expert System," 2017.

# Machine Learning for Process-X: A Taxonomy

Felix Reinhart*, Sebastian von Enzberg, Arno Kühn, and Roman Dumitrescu

Fraunhofer IEM Institute for Mechatronic Systems Design
Zukunftsmeile 1, 33102 Paderborn, Germany

**Abstract.** Application of machine learning techniques for data-driven modeling of value-creating processes promises significant economic benefits. These applications include process monitoring, process configuration, process control and process optimization (process-X). However, similarities and distinguishing features between established approaches to process-X compared to machine learning are often unclear. This paper sheds light on this issue by deriving a taxonomy of process-X approaches that sharpens the role of machine learning in these applications. Moreover, the taxonomy and discussion identifies future research directions for applied machine learning in cyber-physical systems.

**Keywords:** Machine Learning, Data Analytics, Statistical Process Control, Control Theory, Virtual Sensors, Expert systems

## 1 Introduction

Monitoring and prediction of machine conditions based on data-driven models is maturing due to the recent efforts to equip machines with smart maintenance functions. Condition monitoring and predictive maintenance applications aim at reduction of unplanned downtimes and maintenance costs. However, the quality, stability and efficiency of the value-creating processes implemented by these machines often impact a company's profit yields and long-term success even more than increasing plant availability and improving maintenance. Therefore, it is of utmost importance for economic progress to advance existing approaches and develop novel ways to monitor, configure, optimize and ultimately control the processes that are involved in production.

Currently, process monitoring by displaying measured process variables in dashboards, for instance in control charts [1], is an established means to support factory workers in manual process control and configuration. For process optimization, classical model-free and model-based optimization approaches, including design of experiment (DoE, [2]), response surfaces and surrogate methods [3] [4], are state of the art. Furthermore, techniques for process mining and from business intelligence as well as operations research are common to optimize different organizational divisions of a company including production. Process optimization through data analytics is typically implemented by means of off-line analytics-action loops to achieve improvements (see the maturity model in [5]).

---

* Corresponding author

© Springer-Verlag GmbH Germany, part of Springer Nature 2020
J. Beyerer et al. (Eds.), *Machine Learning for Cyber Physical Systems*, Technologien für die intelligente Automation 11,
https://doi.org/10.1007/978-3-662-59084-3_4

In the context of cyber-physical systems, the combination of historical and real-time plant data with data-driven modeling techniques is expected to advance monitoring, configuration, optimization and automated control of value-creating processes beyond such off-line process optimization loops [6]. However, it remains unclear how exactly optimization can be achieved effectively in this context because mostly data for process modeling is very sparse with respect to variation of process parameters, e.g. [7] [8]. Moreover, the lack of solution patterns and indicators renders the design and implementation of future process-X (monitoring, configuration, optimization, and control) functions challenging.

This paper addresses these issues by presenting a taxonomy of established process-X approaches that integrates more recent data-driven schemes for process-X and emphasizes their mutual relations. In particular, data-driven modeling techniques for process monitoring [9] [10] [11], configuration [12], optimization [13] [14] [15], as well as automated process control [16] [17] [18] are included. The taxonomy and further discussion highlights links between traditional approaches from engineering, control theory, stochastic process control, and mathematical optimization to machine learning. Novel process-X approaches leveraging data-driven techniques including current developments in reinforcement learning are integrated into this picture and future developments are anticipated.

## 2    Definition of Process-X

This section introduces the notion of process-X applications that serves as basis for the further discussion:

– **Process Monitoring** is concerned with the measurement, computation and visualization of process variables. Process variables comprise measurements related to the process input, e.g. properties of the feedstock, the process itself, e.g. temperatures and pressures etc., as well as the process output, e.g. quality variables. Primary goals of process monitoring are transparency of process states and the detection of faults, out-of-control states and anomalies. While a process monitor can be an intermediate step towards automated adaptation of process parameters, a distinctive feature of a process monitor is the fact that there is no open or closed loop controller implemented by the monitoring system. Instead the process monitor solely serves as indicator (e.g. by alarming human operators on the shopfloor) for special incidents that causes the process to run out of control (cf. Fig. 1 (a)).
– **Process Configuration** refers to the manual or assisted set-up of production system parameters in particular in the production start-up phase (see Fig. 1 (b)). Process configuration may include the exchange and adaptation of tools, adjusting cycle times, trajectories, as well as setting up reference values for controlled process parameters, e.g. target temperatures, forces, etc. Process configuration can be understood as defining the product- and machine-specific setpoint of the production system, which may be later adopted. Tools for process configuration can also be used for manual process

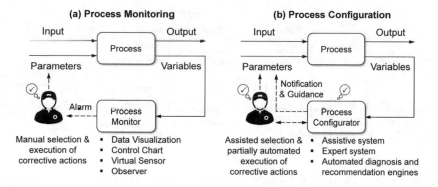

**Fig. 1.** Process Monitoring (a) and Process Configruation (b) schemes with exemplary approaches for implementation.

adaptation in the production phase, e.g. when an incident is indicated by a process monitor.

– **Process Control** aims at maintaining predefined reference values for certain process variables by automatically and continuously adjusting process parameters (see Fig. 2 (a)). For instance, maintaining geometric target features, e.g. angles, in a punch-bending machine by automatically adjusting actuators [19]. We do not include the automatic control of process inputs into this definition, e.g. the control of a temperature in a tank that serves as feedstock for the considered process. Process control can be implemented by closed and open loop controllers.

– **Process Optimization** finally denotes the improvement of predefined criteria connected to a value-creating process (see Fig. 2 (b)). For instance, consider the minimization of scrap and rejected parts by adjusting the setpoint of process parameters. Process optimization can be decomposed into the optimization algorithm, the cost model and the process model.

While the introduced notion for process-X applications shall serve as generic scheme for use cases, the following sections are concerned with the technological implementation of these schemes.

## 3    Machine Learning for Process-X

This section reports examples and schematic patterns for the implementation of process-X applications with a focus on machine learning and data-driven modeling techniques. Under the label of *data-driven*, we subsume all techniques that extract knowledge and construct models as well as representations from data. We group statistical models also under the notion of data-driven modeling. We denote by model-free techniques that are based on humans and simple heuristics, or lack any kind of more complex model or representation (e.g. visualization of raw data or PID control).

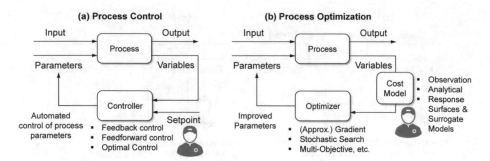

**Fig. 2.** Process Control (a) and Process Optimization (b) schemes with exemplary approaches for implementation.

- **Process Monitoring (cf. Fig. 1 (a)):** Classical approaches to process monitoring include visualization of raw and preprocessed data, including key performance indicators etc. Various flavors of control charts from statistical process control belong to the established techniques in this application domain. Machine learning can be utilized for process monitoring in various ways: Firstly, elaborate data visualizations of high-dimensional and time-series data by means of dimension reduction and embedding techniques are applicable. Secondly, anomaly detection approaches, sometimes also based on the before mentioned approaches, are useful to display process deviations from the norm. Thirdly, trend detection and state transition models can include a predictive capability into the process monitor. Finally, the concept of virtual sensors, also denoted as soft sensors, is suitable to monitor not directly measurable variables [9] [10] [11]. While virtual sensors can be based on analytical models, data-driven learning of virtual sensors may manifest one of the most widely applicable scheme of machine learning to process monitoring. Related to the concept of virtual sensors are observer approaches from control theory [20] [21]. Observers have been utilized for process monitoring and can be also combined with data-driven modeling [18].
- **Process Configuration (cf. Fig. 1 (b)):** Classical approaches for process configuration are based on analytical models of the plant, which provide default setpoints for the process and relations between process input variation and proper process parameters. Based on process models, either analytical or data-driven, also iterative tuning of process parameters may be conducted in a manual manner in interaction with an assistive system for process configuration, e.g. by notifying the worker which variables may achieve the desired outcome with minimum parameter change based on a sensitivity analysis. Machine learning may here be applied for learning process models from historic data, which map product/input-specific variables to process parameters and thus allow the automated extraction of good process parameters. Process configuration in close, physical interaction with the user has been demonstrated in [12], where the user kinesthetically teaches a

compliant robot how to resolve redundancy in order to accomplish a desired task. In this application, the process is configured by the user who provides data for an iteratively trainable inverse kinematics model.

- **Process Control (cf. Fig. 2 (a)):** Process control traditionally is subject to control theory. Machine learning can here mainly serve as substitute for analytical modeling by deriving models from historic and online data. In fact, machine learning has close links to adaptive control and filtering [22]. There is a decent amount of work on feedback error learning and iterative learning control, e.g. [23] [24]. Also, hybrid modeling, where analytical and data-driven modeling approaches are combined [25] [26], is relevant to process control. In particular, the learning of direct inverse models or forward models plus iterative solving by inversion of the locally linearized forward model can be exploited for process control. Current research on autonomous exploration from robotics on learning inverse models, e.g. [27] [28], may unleash the full power of machine learning for process control beyond learning from feedback errors in the future.

  Other relevant fields of research for process control are (stochastic) optimal control and reinforcement learning (RL). Both fields are deeply related [29] [30]. In a simplified notion, we can understand reinforcement learning as data-driven variant of optimal control. The reinforcement learning community does not yet focus much on its possible contribution to process control. A reason may be that the exploration of actions in RL needs to be conducted with special care, because inadequate actions can easily cause high costs e.g. by producing scrap or damaging equipment. It is therefore important to foster analytical, data-driven, or hybrid modeling of production processes for simulation and improvement of control policies. However, data- and sample-efficiency of the modeling still plays a key role also for the modeling.

- **Process Optimization (cf. Fig. 2 (b)):** While process optimization is traditionally based on methods from engineering using mostly analytical models and DoE schemes, data-driven models can substitute or complement analytical models. It is noteworthy that model-based process optimization can be fairly decomposed into the optimization algorithm, the cost model and the process model. While the choice of the optimization algorithm, e.g. (approximate) gradient, stochastic search, etc., is rather less critical, choosing an appropriate approach for process and cost modeling is of utmost importance. Depending on the cost criteria and their complexity, data-driven modeling of the cost function, e.g. as it is done in the response surface and surrogate methodology, is indicated and results in a direct model that maps process parameters to costs from observations. Otherwise, process and costs can be modeled separately, either using data-driven, analytical, or hybrid modeling schemes.

The discussed role of machine learning for process-X applications shows the tight interconnection of data-driven and more traditional approaches. From the arguments above, data-efficiency of the modeling remains central also for machine learning applications. Hence, model complexity needs to be carefully chosen to not over-fit models to parsimonious data. In this context, hybrid modeling and the integration of prior knowledge into the learning process are promising research directions. With restricted amounts of data, the flexibility offered by machine learning models can be turned into improved accuracy with good generalization by exploiting prior knowledge about the process.

# 4 Taxonomy of Process-X Approaches

This section aims at a compact taxonomy of process-X implementation schemes, where we continue with a focus on data-driven modeling techniques. We characterize groups of techniques based on their methodological basis outlined in the previous section. These techniques are labeled with common terms and names of the most relevant research areas. While there may be works that do not entirely fit into this grouping, the taxonomy shall contribute to the understanding and anticipation of possible process-X solutions from a bird's eyes view.

Table 1 distinguishes process-X approaches based on characteristics of the modeling methodology: model-free, analytical, data-driven as well as hybrid modeling (combination of last two columns). In the following, we interpret Table 1 with respect to the roles of the main disciplines concerned with or suitable for process-X applications and anticipate open research and development niches:

– **Engineering:** Engineering has a long history in process-X applications, model-building and process optimization, where constraints like data-efficiency, interpretability and computational efficiency have been major criteria in design and selection of methods. There are interesting historic links between

| Process-X | | Modeling | | |
|---|---|---|---|---|
| | | Model-free / None | Model-based | |
| | | | Analytical | Data-driven |
| Monitoring | | Visualization of raw data | Visualizations, Virtual Sensors, Observer | Visualization, Statistical Process Control, Learned Virtual Sensors |
| Configuration | | Manual configuration by heuristics / experts | Configuration by manual loop with model support, Expert and assistive systems | |
| Control | Process Model | Feedback Control | Feedforward & Feedback Control, Model Predictive Control | Learned Feedforward Control |
| | Cost Model | | (Stochastic) Optimal Control | Reinforcement Learning |
| Optimization | Process Model | Manual, DoE, (Approximate) Gradient, Stochastic Optimization | Model-based Optimization | |
| | Cost Model | | | Response Surfaces, Surrogate Models |

**Table 1.** Taxonomy of Process-X (monitoring, configuration, control, and optimization) approaches with a focus on data-driven modeling.

these methods and machine learning. For instance, the origin of kernel methods can be traced back to the so-called *Kriging* method from geostatistics and meteorology [31]. An example where modern machine learning theory complements classical engineering approaches is the DoE and active learning. While in DoE optimal sampling is already based on information theoretic criteria considering *a priori* information for model building [32], active learning introduces *a posteriori* criteria to select samples while the model is build [33] [32]. A closer integration and broader picture of the history of modeling from an engineering perspective seems to be of high relevance for future teaching in academia in order to establish a deeper understanding of a major application area for machine learning in the future.

– **Control Theory:** Obviously, control theory is mainly relevant to automated process control. However, it is interesting to note that there are concepts from control theory, in particular the concept of observers, that also allows for the implementation of process monitoring applications. An example is [18], where an observer based on an analytical model of a dough kneading process is used to monitor and ultimately terminate the kneading process. Additionally, control theory can also be applied for manual process configuration loops by sensitivity analysis, where a model of the process can serve as basis to derive update recommendations for process parameters.

– **Statistics:** From applied mathematics, statistical methods have mainly driven the field of statistical process control. While statistical approaches share the data-driven methodology with machine learning, the methods are restricted to mostly linear and simpler models. However, statistical approaches must remain in the portfolio of solution patterns for process-X applications and indicators for more complex data-driven modeling techniques, i.e. machine learning, must be defined more clearly.

– **Computer Science:** The field of computer science subsumes expert systems and machine learning. While expert systems have their origin in classical artificial intelligence [34], machine learning may be understood as descendant of the subsymbolic, connectionist branch of artificial intelligence, i.e. neural networks. It is noteworthy that these approaches distribute entirely over rows in Table 1. This underlines the generality and importance of modern computer science in current and future process-X applications and in cyber-physical systems. However, it is also important to note that the acceptance, reliability, and success of machine learning in the domain of engineering requires special care and integration into established practices and tools.

## 5   Conclusion

There is a multitude of machine learning approaches available and even more combinations of these methods possible to implement process-X applications. This makes the conception and exploitation of machine learning for process-X difficult. While the authors already pointed out the difference of analytical and machine learning modeling techniques with respect to the particular challenges and characteristics of their application in cyber-physical systems [26] [35] [36], the relation of machine learning approaches to process-X applications compared to established techniques still remained fuzzy. The taxonomy presented in this paper addressed this issue by relating and structuring possible methodological approaches to process-X applications. Historic links between the different approaches have been highlighted and potential directions for future research have been identified. It will be of utmost importance for efficient and effective development of process-X applications to build upon and integrate existing approaches to process-X with application of machine learning as case-specific requirements indicate. Support of the development process for process-X applications may be achieved in the future by developing morphological boxes and solution patterns for the entities in the presented taxonomy.

# References

1. J.S. Oakland. *Statistical Process Control.* Butterworth-Heinemann, 1986.
2. G.E.P. Box and K.B. Wilson. On the experimental attainment of optimum conditions. *J. R. Stat. Soc.*, 13(1):1–45, 1951.
3. D.R. Jones. A taxonomy of global optimization methods based on response surfaces. *Journal of Global Optimization*, 21(4):345–383, 2001.
4. D.C. Montgomery. *Design and Analysis of Experiments: Response surface method and designs.* John Wiley & Sons, 2005.
5. K. Steenstrup, R.L. Sallam, L. Eriksen, and S.F. Jacobson. Industrial Analytics Revolutionizes Big Data in the Digital Business. Technical report, 2014.
6. K.L. Lueth, C. Patsioura, Z.D. Williams, and Z.Z. Kermani. Industrial Analytics – The current state of data analytics usage in industrial companies, 2016.
7. A. Mangal and N. Kumar. Using big data to enhance the bosch production line performance: A Kaggle challenge. In *IEEE Intern. Conf. on Big Data*, pages 2029–2035, 2016.
8. A. Unger, W. Sextro, S. Althoff, T. Meyer, K. Neumann, R.F. Reinhart, M. Broekelmann, K. Guth, and D. Bolowski. Data-driven Modeling of the Ultrasonic Softening Effect for Robust Copper Wire Bonding. In *Intern. Conf. on Integrated Power Systems*, 2014.
9. C. Atkinson, M. Traver, T. Long, and E. Hanzevack. Predicting smoke. *InTech*, pages 32–35, 2002.
10. F. Gustafsson, M. Drevö, U. Forssell, M. Löfgren, N. Persson, and H. Quicklund. Virtual Sensors of Tire Pressure and Road Friction. Technical report, 2001.
11. L. Fortuna, S. Graziani, A. Rizzo, and M.G. Xibilia. *Soft Sensors for Monitoring and Control of Industrial Processes.* Springer, 2007.
12. S. Wrede, C. Emmerich, R. Grünberg, A. Nordmann, A. Swadzba, and J. Steil. A user study on kinesthetic teaching of redundant robots in task and configuration space. *Journal of Human-Robot Interaction*, 2(1):56–81, 2013.
13. S.K. Lahiri and N.M. Khalfe. Soft sensor development and optimization of a commercial petro-chemical plant integrating support vector regression and genetic algorithm. *Chemical Industry & Chemical Engineering*, 15(3):175–187, 2009.
14. T. Meyer, A. Unger, S. Althoff, W. Sextro, M. Broekelmann, M. Hunstig, and K. Guth. Reliable manufacturing of heavy copper wire bonds using online parameter adaptation. In *IEEE Electronic Components and Technology Conference*, pages 622–628, 2016.

15. W. Aswolinskiy, R.F. Reinhart, and J.J. Steil. Modelling of parametrized processes via regression in the model space of neural networks. *Neurocomputing*, 2017.
16. Z.-S. Hou and Z. Wang. From model-based control to data-driven control: Survey, classification and perspective. *Information Sciences*, 235:3–35, 2013.
17. I. Bratko. Modelling operator's skill by machine learning. In *International Conference on Information Technology Interfaces*, pages 23–30, 2000.
18. F. Oestersötebier, P. Traphöner, R.F. Reinhart, S. Wessels, and A. Trächtler. Design and implementation of intelligent control software for a dough kneader. *Procedia Technology*, 26:473–482, 2016.
19. M. Borzykh, U. Damerow, C. Henke, A. Trächtler, and W. Homberg. Model-Based Approach for Self-correcting Strategy Design for Manufacturing of Small Metal Parts. In *Digital Product and Process Development*, pages 320–329, 2013.
20. J.J.E. Slotine and W. Li. *Applied Nonlinear Control*. Prentice Hall, 1991.
21. Hassan K. Khalil. *Nonlinear Systems*. Macmillan Publishing Company, 1992.
22. K. J. Astrom. Adaptive control around 1960. In *IEEE Conference on Decision and Control*, volume 3, pages 2784–2789 vol.3, 1995.
23. Jun Nakanishi and Stefan Schaal. Feedback error learning and nonlinear adaptive control. *Neural Networks*, 17(10):1453–1465, 2004.
24. D.A. Bristow, M. Tharayil, and A.G. Alleyne. A survey of iterative learning control. *IEEE Control Systems*, 26(3):96–114, 2006.
25. J.F. Queisser, K. Neumann, M. Rolf, R.F. Reinhart, and J.J. Steil. An active compliant control mode for interaction with a pneumatic soft robot. In *IEEE/RSJ Intern. Conf. on Intelligent Robots and Systems*, pages 573–579, 2014.
26. R.F. Reinhart, Z. Shareef, and J.J. Steil. Hybrid analytical and data-driven modeling for feed-forward robot control. *Sensors*, 17(2), 2017.
27. M. Rolf, J.J. Steil, and M. Gienger. Online goal babbling for rapid bootstrapping of inverse models in high dimensions. In *IEEE Intern. Conf. on Development and Learning*, pages 1–8, 2011.
28. R.F. Reinhart. Autonomous exploration of motor skills by skill babbling. *Autonomous Robots*, (7):1–17, 2017.
29. R.S. Sutton, A.G. Barto, and R.J. Williams. Reinforcement learning is direct adaptive optimal control. *IEEE Control Systems*, 12(2):19–22, 1992.
30. S. Bhasin. *Reinforcement Learning and Optimal Control Methods for Uncertain Nonlinear Systems*. PhD thesis, University of Florida, 2011.
31. E.L. Snelson. *Flexible and efficient Gaussian process models for machine learning*. PhD thesis, University College London, 2007.
32. L. Paninski. Design of experiments via information theory. In *Advances in Neural Information Processing Systems*, pages 1319–1326, 2004.
33. D.A. Cohn, Z. Ghahramani, and M.I. Jordan. Active learning with statistical models. *Journal of Artificial Intelligence Research*, (4):129–145, 1996.
34. P. Jackson. *Introduction to expert systems*. Addison-Wesley, 1986.
35. R.F. Reinhart. Industrial Data Science – Data Science in der industriellen Anwendung. *Industrie 4.0 Management*, (32):27–30, 2016.
36. R.F. Reinhart, A. Kühn, and R. Dumitrescu. Schichtenmodell für die Entwicklung von Data Science Anwendungen im Maschinen- und Anlagenbau. In *Wissenschaftsforum Intelligente Technische Systeme*, pages 321–334, 2017.

# Intelligent edge processing

Ljiljana Stojanovic[1]

[1] Fraunhofer IOSB, Fraunhofer Straße 1, 76131 Karlsruhe, Germany
ljiljana.stojanovic@iosb.fraunhofer.de

**Abstract.** Innovating maintenance is crucial for the competitiveness of the European manufacturing, pressured to increase flexibility and efficiency while reducing costs. Initiatives related to Industrie 4.0 have been showing the potential of using advanced/pervasive sensing, big data analytics and cloud-based services. In this paper, we present the edge part of our solution for self-healing manufacturing to early-predict equipment condition and make optimized recommendations for adjustments and maintenance to ensure normal operations. The intelligent edge is advanced, affordable and easily integrated, cyber-physical solution for predicting maintenance of machine tools in varying manufacturing environments, by using new connectivity, sensors and big data analytics methods. The proposed solution is capable to integrate information from many different sources, by including structured, semi-structured and unstructured data. The key innovation is in IoTization through dynamic, multi-modal, smart data gathering and integration based on the semantic technologies.

**Keywords:** Edge Computing, Semantics, Intelligent data fusion

## 1 Introduction

According to [1], the number of smart connected devices will grow beyond 50 billion by 2020 and Internet of Things (IoT) has the potential to represent 11% of the world's economy over the same period. The manufacturing sector is expected to be one of the top adopters of IoT technologies. On the other hand, IoT has been considered as one of disruptive technologies that have the maximum potential for revolutionizing the landscape of manufacturing [2].

Leveraging the power of advanced sensing technologies, applications, such as remote monitoring, anomaly detection, diagnosis, and control of processes and assets have already gained rapid popularity. While huge progress on making assets 'smarter' and production more efficient has been made during last years, the full potential of using Industrial IoT (IIoT) has not yet been exploited sufficiently. The main reasons are:

- the available bandwidth for the data transmission is usually not sufficient for the vast amount and frequency of data created by IIoT devices;
- manufacturing domain has very demanding ultra-low latency requirements for processing IIoT data;
- IoT devices generate wide variety of data formats, building complexity and thus undermining the full benefits of integrated IoT data.

© Springer-Verlag GmbH Germany, part of Springer Nature 2020
J. Beyerer et al. (Eds.), *Machine Learning for Cyber Physical Systems*, Technologien für die intelligente Automation 11,
https://doi.org/10.1007/978-3-662-59084-3_5

To help manufacturing to fully take advantage of IoT, we propose an approach to intelligently handle the IoT sensor data at the edge of the network. The key innovation is in IoTization through dynamic, multi-modal, smart data gathering and integration based on the semantic technologies.

## 2     Our approach for intelligent edge processing

The full potential of using IoT can be achieved by advanced, affordable and easily integrated, edge-based solutions that realizes a holistic, semantic-based approach for dynamic, multi-modal, smart data gathering and integration. Figure 1 summarizes the proposed solution. It provides services to gather sensor data, extract and integrate knowledge, process it locally, and then push it to the cloud.

As shown on Figure 1, we do not focus purely on asset data. We consider data from the surrounding environment in which the asset operates and the associated processes and resources that interact with the asset. Thus, structured data (e.g. machine sensor data), semi-structured data (e.g. inspection reports) and unstructured data (e.g. images), have to be considered by pushing them into the edge or by pulling them from the external systems.

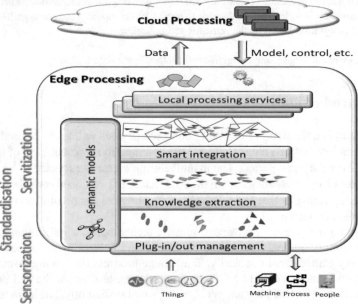

**Fig. 1.** Intelligent edge processing

The intelligent edge is used to connect and manage the equipment in the field to improve availability and support. It is responsible for communicating with the industrial asset, for running local processing (like edge analytics) and for transferring pre-processed data to the cloud. It can be installed either on a machine or on a gateway, depending on the complexity and resource-consumption of services to be used.

Looking at the figure from bottom to top, the components are organized in layers (with push connectors) starting from data sources at the bottom to the communication services[1] that support bidirectional communications between the edge services and the cloud services [6]. The vertical layer services are realized by the semantic models, which provide functionalities that span through all horizontal layer services. They ensure to achieve common, shared understanding across the existing OT and IT systems, across the intelligent edge services as well as across different human roles (e.g. business ana lyst, quality manager, etc.).

The intelligent edge processing is structured in the following sub-layers [6]:

- Plug-in/out management layer enables on demand plug-in/out based on out-of-the-box connectivity. New sensors can be added easily and, depending on the application, some can even be disabled. The main idea is to integrate IoT devices (or even machines) into the existing infrastructure as fast as plugging in an USB device into a PC[2]. This requires capabilities to automatically identify a new or modified device and to integrate it correctly into the running system without manual efforts and changes within the design or implementation of the remaining production system. The proposed approach extends our Plug&Work results [3], which are based on proven industry IEC standards (AutomationML™ and OPC UA). This will accelerate the integration of sensors and devices with the machinery, combine them faster with each other to form a total view on the asset and reduce start-up times.
- Knowledge extraction layer provides support for multi-modal, intelligent sensing. Traditional data analytics works with structured data in well-defined data structures such as numeric and date fields. However, there is a lot of data in semi- and unstructured formats that include multimedia formats such as text, images, video, etc. The use of semi- and un-structured data is a challenge for data analytics in that it has to be pre-processed to the right format before the analytics process [4].

  Indeed, collecting raw sensor data is not enough. It is important to understand the data being analyzed. We propose applying a set of information/knowledge extraction methods to pull out the required information/knowledge from the underlying sources and express it in a structured form suitable for analysis.

  Creating machine-understandable representation of the raw sensor data depends on the data modality. We take into account structured sources (e.g. such machine sensors, vibration sensor, energy sensor, audio sensor, etc.), semi-structured sources (e.g. inspection reports, log data, failure data, etc.), and unstructured sources (e.g. images (corrosion) and video). This requires applying semantic technologies for modelling capabilities of IoT devices and realizing semi-automatic extraction methods to pull out the information/knowledge from the underlying sources. For example, all IoT devices that mimic human sense (i.e. sound (noise detection), sight (camera monitoring), smell (overheated components), touch (heating)) are important for asset health monitoring. The use of advanced technologies provides the necessary

---

[1] E.g. Kafka, MQTT or AMQP
[2] http://www.iosb.fraunhofer.de/servlet/is/51300/

detection sensitivity required to monitor the conditional changes of equipment. The table below shows some examples:

**Table 1.** Examples of ontology-based information extraction

| Data Modality | Approach for extracting semantics |
|---|---|
| Structured data | We generate qualitative and quantitative information by adding (semantic) metadata to content which include not only merging measurement data with the context data (process, machine, worker, location, etc.), but also detecting peaks, sequences, etc. |
| Semi-structured data | Remarks like "Vibration is more than usual" or "Adjust clamps with hitting heads" that can be found in log files, are only human understandable and cannot be processed by software components. By applying shallow NLP based on semantic models, applying sentiment analysis and creating semantic indexes, we are able to reveal detailed insight that is absent in existing solutions. |
| Unstructured data | We extract relevant information from images (e.g. the probability of a certain category like corrosion to be present in an image, the relative size of the object, its spatial context) and represent it as semantic metadata. |

- Smart integration layer ensures data fusion from many sources to uniquely result in a more comprehensive assessment approach. It combines the isolated data in a semantically consistent way, since the intelligence does not come from a single sensor, but from smart connected devices.

We are working on intelligent data fusion in real-time near to the data source. The intelligence lies in dealing not only with syntactic heterogeneity (e.g. different unit of measures for temperature, i.e. degree Celsius or Fahrenheit), but also in semantic heterogeneity (e.g. a gap between assembled parts can be measured manually by a hand-held portable device, specialized equipment system or by a camera). This will significantly improve maintenance processes and will also increase the system reliability, since a faulty device can be semi-automatically "replaced" with another device having the same capabilities.

- Intelligent service layer consists of many services that can be used to build advanced applications suitable for edge. We follow both the model-driven (top-down) approach and the data driven (bottom-up) approach [5]. The model driven approach is based on extracting value from the data sources to find patterns and process complex events (i.e., look for a combination of certain types of events to create a higher-level business event). Here we give an example of patterns: "When an event A arrives at time t and another event B arrives at time t plus x". The data driven approach is based on applying predictive models (that are learnt in the cloud) on the edge. This ensures improving edge analytics over time based on historical analysis.

As shown on the left part of Figure 1, the proposed approach provides contributions to **sensorization** (through multi-modal data gathering) and **servitization** (by applying event-driven architecture on the edge). Additionally, the services have been developed by actively participating in two architecture efforts (**RAMI4.0** and **IIRA**). In the rest of this paper, we focus on the semantics technologies, since they are glue for data, information and knowledge management.

## 3     Role of semantics

The semantic technologies enable formal, machine understandable representation of data. We have applied them to model the different aspects relevant for edge processing: data producers (i.e. sensors and observation), business logic (i.e. domain knowledge), and data consumers (i.e. intelligent services):

- **Sensors and observations**: The data is captured by different hardware, software or human sensors. It is produced by different vendors (or humans in the case of human sensors) and is provided in different formats. One of the biggest challenges in the IoT domain is how to establish the relationship between the raw data and its intended concept. We establish mappings in a semi-automatic way between raw sensors data and the relevant ontologies. For example, the observation data obtained from sensory devices that usually consists of a time stamp stating the time of measurement, device Id, and the values sensed by the sensor, are mapped into the Semantic Sensor Network (SSN) ontology[3]. The advantage of applying semantic technologies to sensor data is conceptualization and abstract representation of the raw data, making it machine interpretable and interlinking the data with existing resources [7].

- **Domain knowledge**: Most real-time data is difficult to interpret without additional domain knowledge. Thus, all available sources (e.g. background knowledge about business processes, strategy, vision, etc.) should be considered. The table below shows some examples for contextualization of raw sensor data.

**Table 2.** Semantic enrichment

| Approach for contextualization | Ontology | Instances |
|---|---|---|
| Mapping of sensor data to tags | Usage of a standard vocabulary or top-level ontologies, e.g. eCl@ss | Information extraction based on data modality; appending statistical aggregates |
| Enriching sensor data with data from relevant software systems | Semi-automatic ontology creation based on database schema | Deep annotation for mapping and migrating legacy data [8] |
| Processing complex events (a combination of certain types of events to create a higher-level event) | Reuse of the ontology for modelling events and patterns [9] | Patterns are manually modelled (e.g. three consecutive measurements, which do not differ by more than 2%) by using concepts and instances introduced above |

---

[3] http://www.w3.org/2005/Incubator/ssn/ssnx/ssn

- **Services**: Within the Industry 4.0 initiative, the Administrative Shell is planned to provide a digital representation of services being available about and from a physical manufacturing component [10]. We combine this approach with the results of the semantic web services[4] and define formal, semantic models for representing services to be executed on the intelligent edge. The semantic description of services allows not only to dynamically add a new service or to replace the existing one, but more importantly to build and verify processing pipelines that consist of several services.

The semantic models have been developed by using the well-known open source tool for ontology creation **Protégé**. Different parsers have been implemented to "**import**" the existing formal models (e.g. an asset model of a machine in AutomationML) and semi-formal models (e.g. a list of failures in a database table) into a common, shared, syntactically and semantically consistent model. The additional challenge (see the 3rd column in the table above) is in **semi-automatic instantiation** of the semantic models. This takes into account both (1) streaming data where data changes over time such as measurement data and (2) data at rest that is fixed (e.g. failure code). The novelty is in **declarative methods** to precisely specify information extraction tasks, and then **optimizing** the execution of these tasks when processing new data.

# 4     Use case

The proposed approach is driven by real industrial use-cases and includes a proof-of-concept demonstration for validation with real-time data. To help test, validate and improve versions of the proposed system, we have developed several industrial-strength smart applications. Here we explain the deployment of the proposed system in the model factory Plug&Work of Fraunhofer IOSB[5] in Karlsruhe.

The demo plant at the Fraunhofer IOSB in Karlsruhe, shown in Figure 3, represents a real production system at a laboratory scale. It consists of 10 modules, which are strictly separated and can be centrally controlled by an MES or can be controlled based on individual logic. The demo plant incorporates a handling facility to fill, transport and empty bottles. The order driven filling is either liquid or granular. Each of the 10 production stations is equipped with a PLC to control the production process. OPC UA servers are used to interface the specific PLC communication to the MES / ERP applications. The factory MES controls the production flow with the planned order sequence. The ProVis.Visu®[6] is used for real-time visualization of production data. The visualization is generated automatically from a factory model in AutomationML[7], a high-level semantic data exchange standard for engineering data.

---

[4] https://en.wikipedia.org/wiki/Semantic_Web_service
[5] https://www.iosb.fraunhofer.de/servlet/is/58621/
[6] https://www.iosb.fraunhofer.de/servlet/is/35793/
[7] https://www.automationml.org

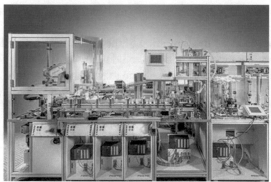

The demo plant is the reference implementation for the recently published OPC UA Companion Specification 'AutomationML for OPC UA' and was used in multiple research projects. Plant models in AutomationML and OPC UA for data communication provide the standards basis.

**Fig. 2.** Demo factory in Karlsruhe

Figure 4 shows the local models defined for each station as well as the global model, which integrates by establishing the mappings among the local models. The demo factory demonstrates Plug & Work scenario showing how new components such as devices or machines can be "plugged" into a factory with an automatic recognition of the component's characteristics and interfaces and an automatic re-configuration of the application functions. Additional quality information provided via MES and/or SCADA to the user can provide information about optimization potential.

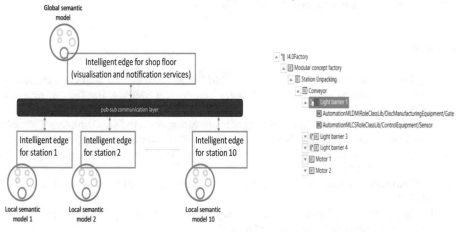

**Fig. 3.** Application of the intelligent edge approach on the demo factory

# 5     Related work

Detailed semantic models should be defined to structure information on the shop floor and support effective use for decision support and cooperation purposes. In this direction, a variety of models has been proposed to encapsulate the shop floor informational structure focusing mainly on factory sustainability, product design and planning, and interoperability. Furthermore, the FP7 project LinkedDesign has proposed a framework for representing industrial information, providing an ontology-based model and rules for its interpretation.

Following the multi-modal approach for the formation of the sensor network, efforts has been made for the semantic annotation of the collected data towards a shop floor modelling framework that puts the prescriptive maintenance related aspects of machinery in its center. In this way, the knowledge base, enabled by its semantics and context-aware intelligent services, will constitute a very useful resource for the design and adaptation of manufacturing processes to increase machine prognostics based on the health state.

# 6    Conclusion

IoT-ization of the shop-floor provides full-fledged machine health monitoring, by a better exploitation of all available data sources. In this paper, we have described an approach to intelligently handle the IoT sensor data at the edge of the network. The approach is based on the usage of semantic technologies for dynamic, multi-modal, smart data gathering and integration. We proposed a three-dimensional semantic model to achieve common, shared understanding across IoT and existing OT/IT systems and discussed different approaches to instantiate it, enrich with associated business rules and represent it in a way to enable reasoning.

# References

1. Big Data in Manufacturing: BDA and IoT Can Optimize Production Lines and the Bottom Line— but Much of the Industry Isn't There Yet, Frost & Sullivan, Big Data & Analytics, December 2016
2. Big Data, XaaS, and IoT Transforming Manufacturing Automation, Disruptive Technologies Transforming Traditional Processes to Enable Smart Manufacturing, July 2016
3. M. Schleipen, et al.: Requirements and concept for Plug&Work. Automatisierungstechnik 63:801-820, 2015
4. D. Riemer, et al, StreamPipes: solving the challenge with semantic stream pipelines. DEBS 2015: 330-331
5. L. Stojanovic, et al., Big-data-driven anomaly detection in industry (4.0): An approach and a case study. BigData 2016: 1647-1652
6. L. Stojanovic, et al., PREMIuM: Big Data Platform for enabling Self-healing Manufacturing, ICE 2017
7. F: Ganz, Automated Semantic Knowledge Acquisition from Sensor Data; IEEE Systems Special Issue, 2016
8. R.Volz, , et al.,: Unveiling the hidden bride: deep annotation for mapping and migrating legacy data to the Semantic Web. J. Web Sem. 1(2): 187-206 (2004)
9. N Stojanovic, et al.,: Semantic Complex Event Reasoning - Beyond Complex Event Processing. Foundations for the Web of Information and Services 2011: 253-279
10. I. Grangel-González, et al.: Towards a Semantic Administrative Shell for Industry 4.0 Components, ICSC, Seite 230-237. IEEE Computer Society, (2016)

# Learned Abstraction: Knowledge Based Concept Learning for Cyber Physical Systems.

Andreas Bunte[1], Peng Li[1], and Oliver Niggemann[1,2]

[1] Ostwestfalen-Lippe University of Applied Sciences, Institut Industrial IT
Langenbruch 6, 32657 Lemgo, Germany
[2] Fraunhofer IOSB-INA, Langenbruch 6, 32657 Lemgo, Germany

**Abstract.** Machine learning techniques have a huge potential to support humans, some impressive results are still achieved, such as AlphaGo. Until now the results are on a sub-symbolic level which is hard to interpret for humans, because we think symbolically. Today, the mapping is typically static which does not satisfy the needs for fast changing CPSs which prohibit the usage of the full machine learning potential. To tackle this challenge, this paper introduces a knowledge based approach of an automatic mapping between the sub-symbolic results of algorithms and their symbolic representation. Clustering is used to detect groups of similar data points which are interpreted as concepts. The information of the clusters are extracted and further classified with the help of an ontology which infers the current operational state. Data from wind turbines is used to evaluate the approach. The achieved results are promising, the system can identify the operational state without an explicit mapping.

**Keywords:** Clustering, Ontology, Knowledge, Reasoning, Classification

## 1    Introduction

Machine learning techniques are common in industrial environments and are used for many tasks, such as anomaly detection, predictive maintenance and optimization. But there is always a gap between the algorithms and operators, because machine learning is performed on the sub-symbolic layer, whereas humans are thinking on a symbolic layer, which means they think in concepts. This can be illustrated at the clustering task, which is an unsupervised machine learning technique. It assigns similar data points to same clusters. Thus, it is comparable to a concept which aggregate things with similar features. The difference between a cluster and a concept is that the cluster is generated sub-symbolically, so that there is no meaning which can be assigned to the cluster. Therefore, the communication between humans and such techniques is difficult, because there is no benefit if the machine shows humans: "The current state is cluster 5". Therefore, an abstraction from a cluster to a concept is needed, which technically means a type assignment of a cluster. This could be performed manually, but since the cyber physical system (CPS) are changing fast, it should be performed automatically.

© Springer-Verlag GmbH Germany, part of Springer Nature 2020
J. Beyerer et al. (Eds.), *Machine Learning for Cyber Physical Systems*, Technologien für die intelligente Automation 11,
https://doi.org/10.1007/978-3-662-59084-3_6

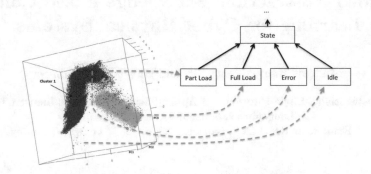

**Fig. 1.** Aim of the work is an automatic assignment of clusters to an operational state

To perform an automatic type assignment (see figure 1) a knowledge based approach is developed. Knowledge is used to interpret the results and especially to describe the states which should be inferred. Therefore, ontologies are used for representation, which define classes, properties and individuals.

The contribution of this paper is a novel approach to assign clusters to a corresponding concept and automatically name them uniquely. This is achieved by using an knowledge-based approach and by reasoning about the knowledge. It bridges the gap between sub-symbolic and symbolic results and thus enables the machine to express results in a human understandable manner.

The paper structures as follows: An overview about the state of the art is given in section 2. Section 3 introduces the approach in detail. The results are presented in section 4 by using a concrete use-case. Section 5 summarize the work.

## 2    State of the Art

In this section a literature review of ontology and concept learning is given. Ontology learning is not the main topic but relevant, because it is similar to this work by adding instances to the ontology and classify them.

In principle the learning of ontologies is possible, but it is almost used for lightweight ontologies such as taxonomies. Hierachichcal clustering or decision trees are typical methods which are used for it, because they can be translated directly into lightweight ontologies. Nevertheless, existing approaches often focus on linguistic properties such as [7] or [6]. That means, structured, semi-structured and unstructured texts are used and similarity measures such as syntax, properties of word or the probability of occurrence are used to derive relations between words [3]. Such approaches are not transferable to CPS, since there are no text bases available.

There is some work in learning more complex ontologies, but these approaches lack of accuracy for a real world applications. For example Zhu [8] used Bayesian networks to learn more complex ontologies. But even if it is one of the best algorithms that can be find in the literature, the F1-Measure is between 0.3 and 0.8. This indicates the difficulty of ontology learning.

Concept learning is the main focus of this paper, which means that patterns in the data should be learned and assigned to concepts. Most of the approaches use data driven methods, which require an interaction. For example, Araki [2] developed an multimodal Latent Dirichlet Allocation (MLDA) algorithm for a robot which enables an online learning. Therefore, 120 objects are classified in 24 categorize. The algorithm learns a word for every object and categorize each object to a class, but there is a human in the loop, which should be avoided in this work. Other approaches are dealing with texts and learn concepts out of it, such as [1]. A more detail review about the concept learning is given by Lake [4] or Mahmoodian [5].

## 3    Approach

To bridge the gap between sub symbolic and symbolic layer, three steps have to be performed: *(i)* information about the clusters have to be extracted *(ii)* information have to be classified into a symbolic representation and named uniquely *(iii)* usage of a reasoner to infer new knowledge. The presented approach is illustrated in figure 2. The mapping unit is a software component which comprises behavior knowledge about the general procedure of the step *(ii)*.

### 3.1    Information Extraction

For more complex CPS, the system behavior might consist of multiple operational states that depend on different factors, e.g. work environments, operations of the systems. For example, a wind turbine has various operational states, such as idle, part load, full load or error. Information about different states of a system is often not recorded during data acquisition. When cluster analysis is performed on a data set, multiple clusters can be recognized. Each cluster corresponds to a particular operational state of the given CPS. In this step, the information

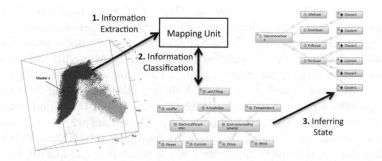

**Fig. 2.** The mapping is done in three steps: information extraction, information classification and inferring

that describes the clusters sub-symbolically will be extracted in the following manner. The names of the variables in the data sets will be passed to the mapping unit, so that the semantic information of each cluster can be retained as much as possible. Commonly, there are different scales of variables in the CPS. To compare the influence of these variables on the mapping task, each variable will be firstly normalized in the range of $[0, 1]$. Furthermore, the minimal values, mean values and maximal values of the variables in a particular cluster will be calculated. This information is the basis for the mapping to a symbolic level.

## 3.2   Classification of Information

In this section it is described how the classification of the cluster information is performed and how to name clusters uniquely and meaningfully. The classification bases on signal values of the cluster center, where every signal value $s \in S$ is assigned to a category $c \in C$ using the function $f$. Five categories $C$ are used to represent the ordinale scale, named no ($\alpha$), low ($\beta$), mean ($\gamma$), high ($\delta$) and very high ($\epsilon$). The number of categories should be chosen according to the use-case, but for many application five categories seems to be suitable.

Because the values of some signals do not use the whole range between 0 and 1, every signal has a parameter to chose the mean value $\mu$. For example, the power factor should be higher than 0,9 and a value near 0 would immediately lead to a blackout. But also the variance can be smaller, e.g. that values are in a range between 0.4 and 0.6. Therefore another parameter $\lambda$ can be set which scales the range down. If there is no parameter defined, the $\mu$ is set to 0.5 and $\lambda$ is set to 1. The boarders for the values are defined as follows:

$$f(s) = \begin{cases} \alpha & \text{if } 0 \leq s \leq \frac{\lambda}{10} \\ \beta & \text{if } \frac{\lambda}{10} < s \leq \mu - \frac{(1-\mu)}{3 \cdot \lambda} \\ \gamma & \text{if } \mu - \frac{(1-\mu)}{3 \cdot \lambda} < s \leq \mu + \frac{(1-\mu)}{3 \cdot \lambda} \\ \delta & \text{if } \mu + \frac{(1-\mu)}{3 \cdot \lambda} < s \leq 1 - \frac{(1-\mu)}{3 \cdot \lambda} \\ \epsilon & \text{if } 1 - \frac{(1-\mu)}{3 \cdot \lambda} < s \leq 1 \end{cases} \tag{1}$$

The formula 1 assigns each signal of a cluster to one concept. This enables to name properties of clusters, such as "The *wind speed* in cluster 3 is *high*." This classification is used to infer the state of the CPS and enables to generate a meaningful and unique name for the cluster. Since every application has some more relevant signals for the process description, so an additional parameter *relevance* can be defined for every signal in the knowledge base. The *relevance* can be set between 0 and 1, whereas 1 means that the signal is mandatory for the cluster description and 0 indicate no relevance of the signal (which will probably occur rarely in practical application). All clusters are described with the mandatory signals, in the first step. They are described by the classified category $c \in C$ and the signal name respectively a name which is defined in the knowledge base. So, a cluster of a CPS could be named with *highPower-lowThroughput* which is obviously not a desired state. If not all clusters are represented with a unique name, more signals are used to describe the ambiguous clusters until their description is unique. The signals are selected regarding their relevance. Thus, the clusters have a unique name which is understandable for humans.

Required knowledge is modeled in an ontology with the web ontology language (OWL). The ontology defines individuals (instances) and concepts (classes) which can be hierarchically and restricted through object and data properties. Individuals are also described through object and data properties. Object properties are describing relations between individuals and data properties are defining values (strings, dates, floats,...) which are used, e.g. for the relevance parameter.

Every cluster is added as an individual to the ontology. The unique formal description is expressed through object properties, e.g. a cluster can be linked to the individual *Power* with the relation *high*. Additionally, data properties are used the store the initial information, namely the values of the cluster center and the name as string. This enables a redefinition of cluster names, e.g. if a new similar cluster is identified.

## 3.3   Reasoning

The main advantage is the classification of instance to clusters, since this is valuable to communicate with operators. The reasoning requires prior knowledge to infer new facts, e.g. to infer the operational state. Possible states have to be defined formally in the knowledge base. The main challenge is to model the knowledge in a way that allows inferences for all possible combinations to exactly one state. Especially, in applications with many states it could lead to wrong definitions. The next section shows exemplary how a state definition looks like.

An OWL reasoner is used to infer knowledge, based on the formal descriptions. Among other things, the reasoner checks all individuals and identifies class assignments for them. Based on the object properties the reasoner checks which cluster fits in which state. The new type assignments is the expected result which should be used to communciate with the operator, the machine is able to provide the information "Currently, the machine is in a error state". Additionally, the unique cluster name can be displayed to the operator with detailed information what kind of error occurred.

## 4   Results

The results for a concrete use-case are presented in this section. Clusters of a wind power plant should be described automatically and determined to an operational state. The data set consists of 12 continuous signals with a time resolution of 10 minutes. Over 230,000 data points generated six clusters, but most of the data is represented in two clusters, (see figure 2).

Information about the maximum, mean and minimum value of every signal in every cluster center is extracted. It is represented as follows:
*Cluster1 windSpeed 0.46 0.20 0.00 rotorSpeed 0.94 0.54 0.00...*
*Cluster2 windSpeed 0.62 0.41 0.22 rotorSpeed 1.00 0.96 0.92...*

At first the values are transformed into a category between *no* and *very high*. This is only performed for the mean value. As described the parameter $\lambda$ and $\mu$ can be used to adapt the categorization. This is done e.g. for the signal *rotorSpeed*, because even at low wind speeds the rotor has a fast drive, the *meanValue* was set to 0.8. The resulting representation is as follows:

*Cluster1 windSpeed* LOW WIND *rotorSpeed* LOW ROTORSPEED...
*Cluster2 windSpeed* MEDIUM WIND *rotorSpeed* VERYHIGH ROTORSPEED...

Then, the cluster descriptions are identified. The signals *wind speed* and *power* are defined as mandatory, since these parameter are important for the operational states and they are interesting for operators. Parameters such as the temperature of the gearbox are less interesting, because they do not correlate strongly with the operational state and they do not represent a common fault. Nevertheless, they are sometimes used, as it can be seen at cluster 1 (in figure 3) which is described as: *low Power - low Wind - medium AirTemperature - medium cosPhi*. This individual holds all information about the cluster in the ontology. The signal values are integrated as object properties, because they might be necessary for determining the concepts, see figure 3.

As last step, the reasoner will assign individuals to classes, in this use-case the operational states should be inferred. Table 1 shows the definition of states, they only base on the two properties wind speed and power. The operational states are not defined very strict, e.g. no wind and low power is defined as idle state, because transitions between two categorizes are always critical. Since two

**Fig. 3.** Example of the individual cluster 1

**Table 1.** Definition of operational states

|  |  | Power | | | | |
|---|---|---|---|---|---|---|
|  |  | no | low | medium | high | very high |
| | no | Idle state | Idle state | Error state | Error state | Error state |
| | low | Idle state | Part load | Part load | Error state | Error state |
| Wind speed | medium | Error state | Part load | Part load | Full load | Error state |
| | high | Error state | Error state | Part load | Full load | Full load |
| | very high | Error state | Error state | Error state | Full load | Full load |

signals are classified, it is not known which one switches first to another category, e.g. if wind and power arise.

The class description of *FullLoad* is shown in figure 4. It is defined as sub class of *OperationalState* and disjoint with the classes *PartLoad, IdleState* and *ErrorState*. Definition of the signals are described by using the *Equivalent To* property, regarding table 1. The yellow background of individual *Cluster4* (in figure 4) indicates that the class assignment was inferred.

The overall results are promising, all clusters are classified correctly. *Cluster 3* is correctly classified as error state, because there is high wind, but no power, so it is obviously for humans that it is an error state. *Cluster 4* is detected as full load state, which is correct, with a mean value of 0.95 for power, the wind power plant is nearly at its rated power.

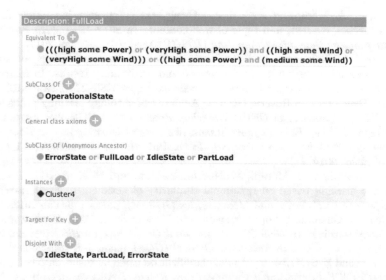

**Fig. 4.** Description of the class full load including one inference

All other clusters (cluster 1, 2, 5 and 6) are classified as part load. They have different combinations of *low/medium* and *power/wind*, additionally the *cos $\varphi$* differs. One would expect that there is an idle state cluster, but the lowest mean value for power of all clusters is 0.095. So, it is at about 10% of the rated power which is obviously no idle state.

## 5 Conclusion

An approach for concept learning is introduced, which maps sub-symbolic data to symbolic information. In a first step a clustering algorithm is performed to generate clusters from data points. These clusters can be interpreted as a concept. Information about the clusters are extracted and classified to one of five categories, namely *no, low, medium, high and very high*. This symbolic representation is added to the ontology. Reasoning is performed in the ontology as a last step. A reasoner assigns the clusters to classes which represent the operational state based on its features. The approach was tested at a wind power plant data set with six clusters. All clusters are assigned correctly to the operational modes.

## Acknowledgment

The work was supported by the German Federal Ministry of Education and Research (BMBF) under the projects "Semantics4Automation" (funding code: 03FH020I3) and "Provenance Analytics" (funding code: 03PSIPT5B).

## References

1. I. Ali, N. A. Madi, and A. Melton. Using text comprehension model for learning concepts, context, and topic of web content. In *2017 IEEE 11th International Conference on Semantic Computing (ICSC)*, pages 101–104, Jan 2017.
2. T. Araki, T. Nakamura, and T. Nagai. Long-term learning of concept and word by robots: Interactive learning framework and preliminary results. In *International Conference on Intelligent Robots and Systems*, pages 2280–2287, Nov 2013.
3. Lucas Drumond and Rosario Girardi. A survey of ontology learning procedures. In *WONTO*, volume 427 of *CEUR Workshop Proceedings*. CEUR-WS.org, 2008.
4. Brenden M Lake. *Towards more human-like concept learning in machines: Compositionality, causality, and learning-to-learn*. PhD thesis, Massachusetts Institute of Technology, 2014.
5. M. Mahmoodian, H. Moradi, M. N. Ahmadabadi, and B. N. Araabi. Hierarchical concept learning based on functional similarity of actions. In *First International Conference on Robotics and Mechatronics (ICRoM)*, pages 1–6, Feb 2013.
6. I. Ocampo-Guzman, I. Lopez-Arevalo, and V. Sosa-Sosa. Data-driven approach for ontology learning. In *2009 6th International Conference on Electrical Engineering, Computing Science and Automatic Control (CCE)*, pages 1–6, Jan 2009.
7. T. Suma and Y. S. K. Swamy. Email classification using adaptive ontologies learning. In *2016 IEEE International Conference on Recent Trends in Electronics, Information Communication Technology (RTEICT)*, pages 2102–2106, May 2016.

8. M. Zhu, Z. Gao, J. Z. Pan, Y. Zhao, Y. Xu, and Z. Quan. Ontology learning from incomplete semantic web data by belnet. In *2013 IEEE 25th International Conference on Tools with Artificial Intelligence*, pages 761–768, Nov 2013.

# Semi-supervised Case-based Reasoning Approach to Alarm Flood Analysis

Marta Fullen[1] and Peter Schüller[2] and Oliver Niggemann[1,3]

[1] Fraunhofer Application Center Industrial Automation, Lemgo, Germany
[2] Marmara University, Istanbul, Turkey
[3] Institute Industrial IT, Lemgo, Germany

**Abstract.** Alarm floods are a major issue in complex industrial plants. Abundance of alarms annunciated in a short period of time can exceed the operators cognitive capabilities and lead to an increased downtime or a serious plant failure. We propose a data-driven approach to detecting and analysing the alarm floods with the goal of supporting the operator during an alarm flood. The approach is based on machine learning concepts of semi-supervised learning and case-based reasoning, and requires a small amount of expert annotations on a historical alarm flood case base. It is comprised of an offline learning stage and an online detection and root cause classification stage. The proposed approach is applied and validated on a real industrial alarm dataset.

## 1 Introduction

Plant operators regularly observe a high number of alarms generated in a short period of time, a phenomenon referred to as alarm flooding. Such a situation poses a threat to the plant operation, likely resulting in a prolonged downtime, increased maintenance costs or even damage to equipment or loss of life [4]. Plant operators are only able to comprehend a limited number of alarms at once, making it difficult to react accordingly. Alarm flooding can have serious and costly consequences and therefore should be carefully addressed in a production plant.

Alarm system design is a process of defining and implementing alarms before the plant begins operation. Its main challenge is avoiding redundancy of alarms and achieving a sufficient degree of robustness that will allow a seamless integration of new plant sections. Unfortunately, it is a difficult process in large and complex industrial plants, especially since it is difficult to foresee the interconnections between the alarms and the future changes in the plant [9].

Thus, alarm systems often show problematic behaviour and generate alarms that are redundant, lingering or chattering. Redundant alarms are registered within different sections of the system, but are triggered by the same factor — and therefore carry no new information. Lingering alarms stay active for a prolonged amount of time, even for days at a time. Therefore they are usually not related to the current situation of the plant. Chattering alarms, on the other hand, are very short and appear in a quick succession — e.g. due to

© Springer-Verlag GmbH Germany, part of Springer Nature 2020
J. Beyerer et al. (Eds.), *Machine Learning for Cyber Physical Systems*, Technologien für die intelligente Automation 11,
https://doi.org/10.1007/978-3-662-59084-3_7

a sensor reading oscillating around a threshold. While chattering alarms do carry a meaningful information, their repeated annunciation disturbs the operator. All those types of alarms clutter the operator interface and make it difficult to correctly assess the plant situation. Moreover, they contribute to increasing the annunciation rate of alarms — the amount of alarms that a plant produces — and a high annunciation rate leads to alarm flooding.

Alarm analysis tools that reduce the number of annunciated and displayed alarms or perform root cause analysis to advise the plant operators can help prevent or reduce the negative consequences, which is crucial in the modern industry. The majority of research in this field is focused on discovering causalities between alarms or building models of the system [7] [8]. Those approaches, however, require a large amount of expert knowledge, which is rarely available. An alternative is to use data-driven approaches, which learn from the available historical data.

We here propose a semi-supervised case-based reasoning approach to support the plant operator in an event of an alarm flood.

Case-based reasoning is used to classify an incoming alarm flood by comparing it to a database of previously known cases. Case-based solutions require a large number of previously seen and annotated alarm floods, but in the reality of industrial plants this is very difficult to obtain. Therefore, we use a semi-supervised approach where the annotations from a small base of floods with known root causes are extended onto the whole set of all previously seen floods.

The approach presented in this work is robust against typical deviations between alarm flood sequences that have the same root cause. Normally, in a large plant, two floods caused by the same root cause will not be exactly identical — e.g. there will be delays in some alarm annunciations or irrelevant alarms will be annunciated during the flood due to other reasons. Here, we consider a distance metric to quantify the similarity between two floods between 0 and 1, and use this metric to find matching floods in the case-base.

The presented approach detects and categorizes floods when they occur, therefore it is easy to apply in practice because it does not require an overhaul of the alarm annunciation system or meddling in the alarm design. The benefit of the system is giving a suggestion to the operator regarding the current situation of the plant (floods are detected on-line and checked whether this type of the flood has been seen previously) and giving a suggestion for the course of action to undertake (the root cause and/or repair instructions of the flood found in the case-base). The approach is intended to classify floods in real-time and support the operator as soon as an alarm flood is detected.

## 2 Preliminaries

### 2.1 Alarm floods in CPSs

In a Cyber-Physical System, an *Alarm* is a message raised by an industrial control system that warns the operator about an unusual condition in the production plant. In the most trivial case an alarm definition is based on a sensor reading exceeding a set threshold.

The *alarm log* is a list of alarms recorded during plant operation. At the very least, it contains an alarm ID, date and time when the alarm was triggered and date and time when the alarm was deactivated. An alarm is deactivated either when the alarm condition is no longer true or the operator manually acknowledges it.

An *alarm flood* is a situation where a high number of alarms is annunciated in a short time. According to the industrial standards [5], an alarm flood starts when more than 10 alarms are annunciated in a 10 minute period and ends when the alarm rate drops below 5 alarms per 10 minutes.

## 2.2   Semi-supervised Machine Learning Methods

*Clustering* is an unsupervised machine learning method that helps discover the underlying, or natural, structure of a collection of entities. It assigns the observations into groups, called clusters, based on the distance between them in such a way that each cluster contain observations highly similar to each other, while dissimilar to observations in other clusters.

An alarm flood *distance measure* is a function $d(f_1, f_2)$, where $f_1$ and $f_2$ are alarm floods. It returns a real number in range $0 \leq d \leq 1$ representing the distance, or the degree of dissimilarity, between the two floods.

*Case-based reasoning* is a machine learning approach for classification of unknown observations based on the comparison with a database of already known observations — "cases". In this work we will apply a simple case-based classification based on a $k$-nearest neighbour principle using a clustering distance function.

*Label propagation* is a semi-supervised machine learning method that allows projecting partial information onto the rest of the dataset [10]. Similar to clustering and case-based reasoning, it is useful for annotating unknown observations based on their similarity to known samples.

## 3   Case-based Approach for Root Cause Classification

Although alarm floods are dangerous and difficult to deal with, it is also difficult or impossible to prevent them since they are usually an effect of a poorly designed alarm system. Therefore we present an approach that supports the operator during a flood, for example by reducing the number of alarms displayed, highlighting critical alarms or providing a suggestion to the cause of the alarm flood.

Our approach, shown in Figure 1, is based on semi-supervised case-based reasoning and comprises two stages: (i) off-line learning, where the historical data is collected and analysed, and (ii) on-line root cause classification, where the results of stage (i) are used to analyse the current situation in the plant.

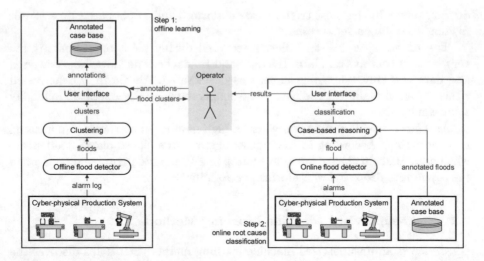

**Fig. 1.** Semi-supervised case-based reasoning approach to alarm flood analysis: offline learning (step 1, left) and online root cause classification(step 2, right).

### 3.1  Step 1: Offline Learning

In this step, historical alarm data is analyzed to prepare a case base for subsequent online case-based reasoning.

The historical alarm logs are preprocessed to make them fit for machine learning. This includes standard preprocessing steps, such as unifying the data structure as well as preprocessing steps specific to alarm analysis: removal of unwanted entries such as warnings, events, lingering alarms, as well as merging chattering alarms.

Alarm floods are detected according to the industry standards: an alarm flood is a period of time where the alarm rate is higher than the alarm flood threshold of 10 alarms per 10 minutes. The flood is considered to begin when the alarm rate first exceeds the threshold, and to end when the alarm rate drops below 5 alarms per 10 minutes. A flood includes all the alarms that were active during this time period, excluding those that began too long before and have not been removed during preprocessing.

Preprocessing and flood detection is performed by an offline flood detector and the outcome is a list of floods identified in the alarm log.

Clustering of alarm floods is used to gain insights into the historical data. The outcome of this step are groups of similar floods, as well as a group of floods that were outliers — they not similar to any other flood.

Finally, results of the analysis are displayed to a human expert who will assess their correctness and provide, at least partial, an annotation for the detected floods — the root cause for the failure that triggered the alarm flood. Floods that are considered similar from the machine learning point of view can be assumed to be caused by the same root cause [1].

Analysing and annotating clusters containing recurrent, similar floods is also easier than analysing floods one by one. The partial annotation can be propagated onto the whole dataset, for example using Label Propagation (Section 2.2), although it is expected that a part of the floods will remain not annotated.

The final result of the off-line learning step is a partially annotated case base of alarm floods.

## 3.2   Step 2: Online Root Cause Classification

During plant operation, our approach uses the case base and an online flood detection mechanism together with all other components from Step 1, to perform online alarm flood classification for supporting a human operator.

If a flood is detected, case-based reasoning is immediately performed, even though the extent and sequence of alarms in the flood is not yet known (as the flood has just started and did not finish yet). To advise the operator, case-based reasoning compares the currently known prefix of the flood to the case base using flood-flood distance metrics. The most similar previously known flood is found and, if the confidence level of the comparison is high enough and the similar flood has an annotation, this annotation is suggested to the operator as a potential root cause. If no such flood is found, case-based reasoning is repeated at a later time when more information about the flood is available.

If the flood ends and no similar flood was found, this flood is added to the case base and awaits annotation from experts for use in the future.

## 4   Experimental setup

We test our approach using a real industrial alarm log. The alarm log contains over 50 thousand alarm annunciations of over 2 thousand different alarms and spans over half a year of plant operation. The alarm log is preprocessed to remove invalid entries as well as merge chattering alarms. Partial annotations from domain experts of the plant operator are available.

### 4.1   Distance Measure and Flood Representation

The distance measure plays a leading role in this approach since it is used at multiple steps of analysis. Multiple existing distance measures can be employed to compare alarm floods [1, 3]. A choice of the distance measure is tightly coupled with choosing an appropriate flood representation. We here use Euclidean distance of TF-IDF representation as this performed well in previous experiments.

TF-IDF (term frequency - inverse document frequency) is a data representation commonly used in the field of natural language processing [6]. In the case of alarm floods, a single alarm corresponds to a term and the whole flood is analogous to a document. This type of representation gives particular importance to alarms which appear rarely and therefore carry the most discriminative information.

TF-IDF is defined for each alarm $a$ and flood $f$ as

$$\text{tf-idf}(a, f) = \text{tf}(a, f) * \text{idf}(a). \tag{1}$$

Term frequency is defined as

$$\text{tf}(a, f) = \frac{f_{a,f}}{|f|}, \tag{2}$$

where $f_{a,f}$ is the number of annunciations of alarm $a$ in flood $f$ and $|f|$ is the total number of annunciations in $f$. Inverse document frequency is defined as

$$\text{idf}(a) = \log_e \frac{|F|}{|\{f \in F \mid a \in f\}|}, \tag{3}$$

where $|F|$ is the total number of floods. The TF-IDF flood representation is a 1-dimensional vector of length $m$, the total number of unique alarm IDs. On this representation we use Euclidean distance, defined for two floods, $f_i$ and $f_j$, as

$$d(f_i, f_j) = \sqrt{\sum_{k=1}^{m} (\text{tf-idf}(a_k, f_i) - \text{tf-idf}(a_k, f_j))^2}. \tag{4}$$

## 4.2   Offline Learning

Offline learning is performed first to construct the case base.

Flood detection is performed over the whole alarm log using a sliding window, and each flood is recorded as a triple: flood start timestamp $t_s$, flood end timestamp $t_e$ and the list of alarms $a_0, a_1, ..., a_n$ belonging to it. The flood start and end timestamps are calculated according to the definition in Section 2.1 and a flood includes all the alarms that were triggered in its duration. To avoid including irrelevant lingering alarms, but to account for the alarms that were triggered before the flood started but might be relevant, a flood will also include alarms that were triggered up to $t_l = 60$ minutes before the flood start.

After transformation into TF-IDF representation, floods are clustered using DBSCAN clustering algorithm [2] with parameter $\varepsilon = 0.45$ and Euclidean distance measure. Resulting clusters are analysed by an expert who provides partial annotations regarding the root causes of some of the floods.

Partial annotation is spread over the remaining floods using Label Propagation algorithm. A probability threshold is used to avoid forcing annotations on unrelated observations, so a part of the floods remains unannotated. The resulting set of floods and annotations makes up the case base used in the online stage.

## 4.3   Simulation of Online Root Cause Classification

To assess the performance of our approach, we simulate online root cause classification of a plant.

Simulation is based on prefixes of floods that are not in the case base. Such a prefix is a subsequence of alarms belonging to a flood, starting at the flood start $t_s$ and ending at an arbitrary point of time $t_p$, up to the flood end $t_e$. The prefix contains all the alarms annunciated between $t_s - t_l$ and $t_p$.

In the following experiments, for each flood, a new prefix is generated for every minute between $t_s + 10$ (when the flood is first detected) and $t_e$ to simulate online flood detection; prefix $p_0$ corresponds to the time period between $t_s$ and $t_p = t_s + 10$, prefix $p_1$ corresponds to the time period between $t_s$ and $t_p = t_s + 11$ and so on, up till $p_{max}$ which corresponds to the time period between $t_s$ and $t_p = t_e$ for the given flood. Due to the industrial definition of alarm flood, therfore a flood prefix is always 10 minutes long and the number of prefixes we consider for a flood depends on its duration in minutes.

Each generated prefix is transformed into TF-IDF representation and processed by the case-based reasoning algorithm, which classifies it using 1-nearest neighbour approach with Euclidean distance measure.

### 4.4    Assessing the results

To assess the performance of our approach, we define three accuracy measures $a_1, a_2, a_3$ based on the standard accuracy measure in supervised machine learning: $a = \frac{n_s}{n_k}$, where $n_s$ denotes the number of correct predictions and $n_k$ denotes the overall number of predictions.

Accuracy measure $a_1$ represents the average of classification correctness when the flood is first detected and presented to the operator, so at prefix $p_0$. It represents the fraction of the floods that were immediately correctly classified, even though only partial information was available.

Accuracy measure $a_2$ represents the average classification correctness when the flood is over, so at prefix $p_{max}$ (the length of which varies from flood to flood). This accuracy measure represents the fraction of the floods that were classified correctly at their end, so based on the full information.

Accuracy measure $a_3[i]$ measures classification correctness after $10 + i$ minutes. It is calculated as an average accuracy of classifications for prefix $p_i$ among all the floods of length equal to or longer than $p_i$.

## 5    Results

Figure 2 shows the number of alarms annunciated in the alarm log, summarized over subsequent 10-minute periods. In the worst case, more than 350 alarms were triggered in a given ten-minute period. This shows that alarm flooding is a real problem and needs to be addressed.

We performed a simulation of online root cause classification on the dataset and obtained an average accuracy at prefix $p_0$ of $a_1 = 0.61$ and an average accuracy at prefix $p_{max}$ of $a_2 = 0.76$. Predictably, the accuracy of classification is higher when it is performed on the longest flood prefix containing the full information.

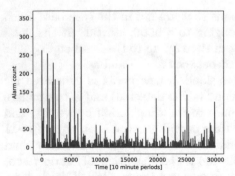
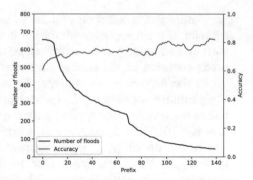

**Fig. 2.** Number of alarms annunciated in each 10-minute period recorded within the alarm log.

**Fig. 3.** Accuracy $a_3$ and number of floods up to a prefix length of 150 minutes.

Figure 3 shows the average classification accuracy for each prefix length $a_3$ along with the number of floods that actually contain each prefix. At prefix $p_0$, of 10 minutes length, all the floods are considered and therefore the accuracy $a_3[0]$ is equal to $a_1$. At any subsequent prefix $p_i$, of $i + 10$ minutes in length, we calculate accuracy $a_3$ only among those floods that contain it. Predictably, $a_3$ becomes higher as the prefix length increases. We omit prefix lengths above 140 because only few floods fall into that region, hence accuracy results have limited value. Interestingly, accuracy rises rapidly between prefixes $p_0$ and $p_{10}$ and then becomes nearly constant, until the floods become very long (from prefix $p_{90}$ which corresponds to floods that last for at least 100 minutes).

## 6   Conclusions

This work presented a semi-supervised case-based reasoning approach to alarm flood root cause classification. The approach has been applied to a real industrial alarm log to assess its performance, and a real-time alarm flood detection was simulated on the basis of flood prefixes.

The approach presented here shows the core idea of alarm flood analysis and opens up many avenues for future research to improve the results. Firstly, other distance measures should be investigated. Secondly, obtaining the annotations in a reasonable way is a challenge (since it requires manual expert work) as well as handling the unannotated floods in the case base. Finally, measuring how well the algorithm is performing is not straightforward. Three accuracy measures were presented in this work to assess the algorithm performance. However, we are interested in analysing the behaviour of the classification result over time. In the best case, the classification is correct at the very moment the flood is detected and remains the same until the flood is over. In some cases though, the classification result changes over time (once or multiple times) as the new information is added. Defining an accuracy measure to quantify the stability of classification result is planned future work.

Nevertheless, the classification results are promising. Average accuracy measures $a_1$ and $a_2$ show that the root cause classification is more accurate, the more information about a flood is available, with an average accuracy reaching over 0.8 in the best case. Average accuracy at the last prefix $p_{max}$ of a flood is roughly 23% higher than at the first prefix $p_0$. However, accuracy measure $a_3$ suggests that first ten minutes after the flood is detected have the biggest impact on classification accuracy. This observation is important for the practical application of the approach, as it suggests that the root cause of the flood can be detected fairly quickly after the flood begins.

# 7    Acknowledgement

This project has received funding from the European Union's Horizon 2020 research and innovation programme under grant agreement No. 678867.

# References

1. K. Ahmed, I. Izadi, T. Chen, D. Joe, and T. Burton. Similarity analysis of industrial alarm flood data. In *IEEE Transactions on Automation Science and Engineering*, Apr 2013.
2. M. Ester, H.-P. Kriegel, J. Sander, and X. Xu. A density-based algorithm for discovering clusters in large spatial databases with noise. pages 226–231. AAAI Press, 1996.
3. M. Fullen, P. Schüller, and O. Niggemann. Defining and validating similarity measures for industrial alarm flood analysis. In *IEEE 15th International Conference on Industrial Informatics (INDIN)*, July 2017.
4. Health and S. E. (HSE). *The Explosion and Fires at the Texaco Refinery, Milford Haven, 24 July 1994 (Incident Report)*. HSE Books, 1997.
5. Instrumentation, Systems, and Automation Society. *ANSI/ISA-18.2-2009: Management of Alarm Systems for the Process Industries*, 2009.
6. K. S. Jones. A statistical interpretation of term specificity and its application in retrieval. *Journal of Documentation*, 28:11–21, 1972.
7. O. Niggemann and V. Lohweg. On the diagnosis of cyber-physical production systems: State-of-the-art and research agenda. In *Proc. AAAI*, pages 4119–4126. AAAI Press, 2015.
8. B. Vogel-Heuser, D. Schütz, and J. Folmer. Criteria-based alarm flood pattern recognition using historical data from automated production systems (aps). *Mechatronics*, 31:89 – 100, 2015.
9. J. Wang, F. Yang, T. Chen, and S. L. Shah. An overview of industrial alarm systems: Main causes for alarm overloading, research status, and open problems. *IEEE Transactions on Automation Science and Engineering*, 13(2):1045–1061, April 2016.
10. X. Zhu and Z. Ghahramani. Learning from labeled and unlabeled data with label propagation. Technical report, 2002.

# Verstehen von Maschinenverhalten mit Hilfe von Machine Learning

Heinrich Warkentin*, Meike Wocken** und Alexander Maier***

*CLAAS E-Systems KGaA mbH & Co KG,
Bäckerkamp 19, 33330 Gutersloh
**CLAAS KGaA mbH,
Mühlenwinkel 1, 33428 Harsewinkel
http://www.claas.com/
***Fraunhofer-Anwendungszentrum Industrial Automation (IOSB-INA),
Langenbruch 6, 32657 Lemgo
http://www.iosb-ina.fraunhofer.de
{heinrich.warkentin,meike.wocken}@claas.com,
alexander.maier@iosb-ina.fraunhofer.de

**Zusammenfassung.** Das Ziel der Untersuchung ist die Beschreibung des Normalverhaltens einer Erntemaschine, um anomales Verhalten als Abweichung dazu identifizieren zu können. Fokus der Betrachtung liegt auf der Motorüberlastung. Die Methode der selbstorganisierenden Karte liefert Hinweise, dass es für eine Maschine im Druschprozess vier Zustände gibt. Diese vier Zustände und angenommene strukturelle Änderungen hinsichtlich des zugrundeliegenden Datengenerierungsprozess der Motordrehzahl wird mit einem Markov-Switching Modell geschätzt. Die geschätzten Parameter spiegeln die identifizierten Zustände wieder. Es wird auch ein Zustand erkannt, in dem deutlich mehr Alarmmeldungen zur Motorüberlastung auftreten als erwartet, aber die Zustände scheinen rund um die Alarmmeldungen nicht stabil zu sein, sodass weitere Untersuchungen folgen müssen, um eine Vorhersage der Alarmmeldungen zu ermöglichen.

**Schlüsselwörter:** Zustandsüberwachung, Predictive Alarming, Landwirtschaft, Maschinelles Lernen, SOM, Markov-Switching

## 1 Einleitung

Moderne Erntemaschinen sind im Laufe ihrer Einsatzzeit variablen Umweltbedingungen (Boden, Feuchtigkeit, Fruchtart, etc.) ausgesetzt. Um mit den wechselnden Bedingungen umzugehen, sind die Komponenten solcher Maschinen flexibel einstellbar. Diese Einstellungen werden vom Maschinenbediener oder Assistenzsystem vorgenommen, um die möglichst effektiv (geringe Verluste) und effizient (geringe Kosten) zu ernten. Dabei wird eine immer größer werdende Menge von Daten erzeugt.

Im Rahmen des vom BMBF geförderten Forschungsprojekts AGATA (Analyse großer Datenmengen in Verarbeitungsprozessen) sind CAN Bus Daten von

CLAAS Mähdreschern (Typ Lexion) mit dem Ziel der Anomalieerkennung aufgezeichnet worden. Ein Ansatz für die Zustandsüberwachung und das Auslösen von Alarmmeldungen basiert auf Schwellwerten. Daneben gibt es weitere Ansätze, die auch Methoden des maschinellen Lernens verwenden.

Die Abbildung 1 zeigt den Gutfluss des Ernteguts innerhalb eines Lexion schematisch. Die geernteten Pflanzen werden vom Schneidwerk in die Dreschtrommel befördert. Nach dem Druschvorgang wird das Korn weiter vom Rest des Ernteguts separiert, anschließend gereinigt und bei größeren Bestandteilen wieder auf die Dreschtrommel (Überkehr) oder auf den Korntank befördert.

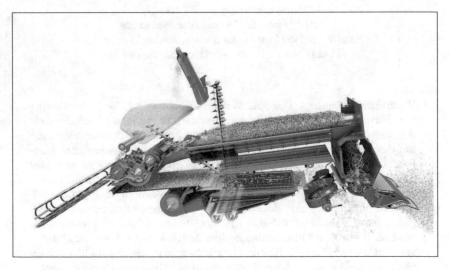

**Abb. 1.** Gutfluss im Mähdrescher (Typ Lexion).

Der Fokus dieser Untersuchung liegt auf der Alarmmeldung zur Überlastung des Motors. Diese führt zu Leistungseinbußen im Ernteprozess. Das Ziel ist eine Beschreibung des Normalverhaltens der Maschine, um anomales Verhalten im Vorfeld des Alarms potentiell identifizieren zu können. Die Maschinenparameter sind jedoch stark von externen Einflussgrößen beeinflusst, die sich nicht direkt in den Daten wiederfinden. Dazu gehören z. B. Ertrag, Feuchtigkeit oder Abreifegrad. Klassische modellbasierte Verfahren oder Zeitreihenanalysen, die die Realität vereinfacht modellieren, bedürfen also Anpassungen.

Eine weitere Herausforderung ist, dass die Maschinendaten nicht immer direkt den Status im Ernteprozess wiedergeben. Ein Beispiel ist die Durchsatzmessung, die am Ende des Gutfluss am Kornelevator stattfindet. Darüber hinaus ist der Druschprozess innerhalb der Maschine teils rekursiv, sodass die Durchsatzmessung einen zeitlichen Versatz enthält. Schwellwertbasierte Alarmmeldungen sind binäre Indikatoren für grenzwertiges Maschinenverhalten. Um Alarmmeldungen vorherzusagen, muss der Maschinenkontext berücksichtigt werden, d. h.

nicht nur der Zustand im Ernteprozess, sondern auch subtilere Veränderungen sind einzubeziehen.

Um den Zustand „Ernten" detaillierter zu untersuchen, soll dieser mithilfe unüberwachter Verfahren analysiert werden. Dazu wird eine selbstorganisierende Karte (SOM, engl. self organizing map) angelegt und aus fachlicher Sicht durch einen Experten analysiert. Diese Ergebnisse werden in einem zweiten Schritt für eine Zeitreihenanalyse verwendet, die die Zusammenhänge in den Teilzuständen beschreiben soll. Für die Zeitreihenanalyse wird ein Markov-Switching Modell genutzt. Auch diese Ergebnisse werden wieder aus Domänensicht analysiert.

Für die Analyse werden Daten eines CLAAS Mähdreschers, die im Rahmen des AGATA Projekts erhoben worden sind, ausgewertet. Die Abtastrate der vorliegenden Maschinendaten beträgt eine Sekunde. Aufgrund unseres Fokus auf den Ernte-Prozess berücksichtigen wir nur Beobachtungen von einer auf dem Feld arbeitenden Maschine, die Weizen erntet.

Unser Vorgehen fassen wir in einem Fazit zusammen. Dabei wird es kritisch reflektiert. Ein Ausblick auf mögliche Erweiterungen der Untersuchung schließt den Beitrag ab.

## 2 Unüberwachte strukturerkennende Analyse mithilfe Selbstorganisierender Karten

### 2.1 Selbstorganisierende Karten

Eine allgemeine Herausforderung für die Datenanalyse ist die Identifikation des Verhaltensmodells technischer Systeme. Dies erfolgt in der Regel unüberwacht. Dazu können verschiedene Methoden verwendet werden. Besonders beliebt hierfür sind Dimensionsreduktionsmethoden wie Principal Component Analysis (PCA) [3] und Self-Organizing Map (SOM). Die SOM basiert auf neuronalen Netzwerken und stellt eine zweidimensionale Karte von Neuronen dar, welche Objekte mit ähnlichen Werten gleichen oder benachbarten Neuronen zuweist. Die Vorgehensweise orientiert sich dabei am menschlichen Gehirn, wobei mehrdimensionale Einganssignale auf eine planare Struktur gemappt werden. In diesem Beitrag wird nur oberflächlich auf den formalen Hintergrund und das Lernen eingegangen. Zur weiteren Vertiefung sei auf die Literatur verwiesen. [4]

In der Lernphase wird eine Menge an Objekten verwendet, um mittels Distanzmetriken die Ähnlichkeit zu den Neuronen der Karte zu berechnen und für jede Beobachtung das Neuron mit dem geringsten Abstand auszuwählen. Anschliessend werden die Abstände des ausgewählten Neurons zu den benachbarten Neuronen angepasst. Dies erfolgt in vielen Iterationen, bis der Lernvorgang konvergiert oder eine vorher eingestellte Menge an Iterationen erreicht wird.

Die gelernte Karte kann mittels Methoden des Visual Analytics verwendet werden, um die Verhaltensstruktur des zugrundeliegenden Systems zu explorieren. Dazu wird die Karte verwendet und für jede Eingangsdimension ein separater Plot erstellt, in dem der durchschnittliche Wert aller Objekte, die einem

Neuron zugewiesen wurden, anhand einer Skala den Farbwert bestimmt. Werden diese Plots der Karten miteinander verglichen, lassen sich Schlüsse über Zusammenhänge zwischen den einzelnen Merkmalen der Eingangsdaten ziehen.

### 2.2  Anwendung und Analyse der SOMs

Zur Auswertung der gelernten Karten wurde zunächst ein Ansatz des Visual Analytics gewählt. Dazu wird für jede der Eingangsdimensionen in einem gesonderten Plot dargestellt. In jedem der parallel dargestellten Plots wird die gleiche Karte verwendet und anhand einer Skala gemäß der Werte aller Objekte, die einem Neuron zugeordnet wurden, eingefärbt.

Die gemeinsame Betrachtung der Plots der einzelnen Dimensionen ermöglicht die Beantwortung von einigen typischen Fragestellungen:

- **Identifikation von Systemzuständen**
  Zur Identifikation von Systemzuständen können Clusteringverfahren (z.B. k-means) genutzt werden. Des Weiteren können die Cluster auch mittels Expertenwissen ermittelt werden.
- **Analyse der Wirkungszusammenhänge im System**
  Basierend auf den identifizierten Clustern werden die Wirkungszusammenhänge im System weiter analysiert, indem die parallelen Plots evaluiert werden.
- **Analyse von Alarmen und deren Umfeld**
  Ähnlich zu der Analyse der Wirkungszusammenhänge im System können nun spezifischere Fragestellungen betrachtet werden. Einen besonderen Stellenwert nimmt dabei die Analyse von Alarmen und deren Umfeld ein. Im Falle des Auftretens eines Alarms ist die Frage nach der Ursache von großer Bedeutung. Die gelernte Karte zeigt den genauen Ort des Auftretens des Alarmes auf. In Kombination mit den Plots der anderen Dimensionen lässt sich auf die regelmäßigen Umgebungsbedingungen schließen, unter denen der Alarm auftritt.

Die Abbildung 2 zeigt eine gelernte SOM im Ernteprozess eines Mähdreschers. Es sind 4 grundsätzliche Phasen zu erkennen:

1. (zu) hohe Belastung
2. (zu) geringe Belastung
3. Gute Leistung mit hohem Durchsatz
4. Gemischte Leistung mit teils höheren Verlusten

Des Weiteren ist ein Bereich erkennbar, in dem gehäuft Alarme aufgrund zu hoher Motorlast auftreten. Die identifizierten Zustände werden im nächsten Abschnitt für weitere Analysen verwendet.

## 3   Zeitreihenanalyse bei unbeobachtbaren Zuständen

Die Zeitreihenanalyse ist ein wichtiges Instrument, um die Beziehungen beim Entstehen einer Zeitreihe zu verstehen und später auch kontrollieren zu können.

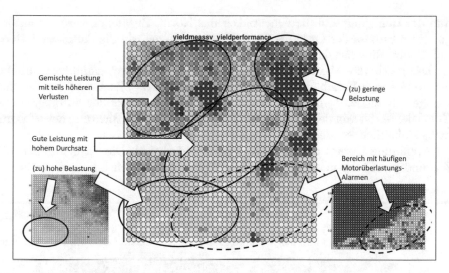

**Abb. 2.** Identifizierte SOM mit den ermittelten 4 Zuständen im Druschprozess und dem Alarm "Motorüberlast".

Aufgrund unseres besonderen Interesses an der Alarmmeldung zur Überlastung des Motors, versuchen wir die Motordrehzahl mit einem Zeitreihenmodell zu erklären.

Die SOM liefert Hinweise darauf, dass wir vier Zustände für die Maschine im Druschprozess annehmen sollten. Die Zeitreihe der Motordrehzahl, in Abbildung 3 zu sehen, scheint zustandsbedingt zu sein: es gibt Phasen unterschiedlicher Varianz der Motordrehzahl. Wir gehen von sich wiederholenden strukturellen Änderungen im Entstehungsprozess der Zeitreihe aus und nehmen an, dass die Parameter des zugrundeliegenden Datengenerierungsprozesses sich von Zustand zu Zustand ändern.

## 3.1 Markov-Switching Modell

Das Markov-Switching Modell, eingeführt durch [1],

$$y_t = \mu(s_t) + \sigma(s_t)\varepsilon_t, \varepsilon_t \sim N(0,1) \tag{1}$$

ermöglicht es, für $s_t \in (1,2,\ldots,k)$, d. h. für $k$ unterschiedliche Zustände flexibel eine Modellierung der Zusammenhänge durchzuführen.

Es wird angenommen, dass die Zustandsvariable $s_t$ einem Markov-Prozess folgt.

Die Schätzung wurde mit der Software-Umgebung R und dem R-Paket MSwM durchgeführt. Als erklärende Variablen verwenden wir Durchsatz, Soll-Werte von Maschineneinstellungen (z.B. Drehzahlen) und die Abweichungen zu den Soll-Werten, den Neigungswinkel der Maschine, die Korn-Verluste in der Separierung und der Reinigung und die Feuchtigkeit des Korns. Zusätzlich wird ein autoregressiver Prozess der Ordnung 10 mit aufgenommen, da die Autokorrelationsfunktionen Hinweise darauf geben.

Das geschätzte Markov-Switching Modell hat statistisch signifikante Effekte der erklärenden Variablen in den Zuständen 1, 3, und 4. Zustand 2 hat deutlich weniger statistisch signifikante Einflussgrößen. Der Effekt der autoregressiven Koeffizienten ist von der statistischen Signifikanz und der Richtung der Wirkung nicht einheitlich, sondern unterscheidet sich in den Prozessen.

Abbildung 3 zeigt den Verlauf der Motordrehzahl. Die Farben geben den Zustand gemäß der maximal gefilterten Wahrscheinlichkeit an.

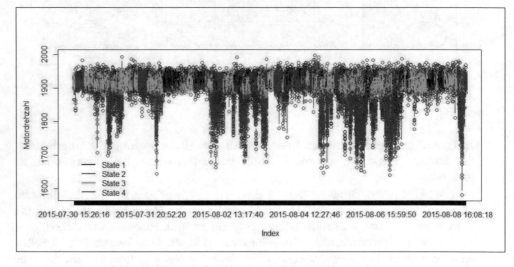

**Abb. 3.** Motordrehzahl mit geschätzten Zuständen.

## 3.2    Analyse Modellschätzung

Ein Ergebnis der Modellschätzung ist eine Matrix mit den Wechselwahrscheinlichkeiten (siehe Tabelle 1) von einem Zustand in den nächsten Zustand. Die Auswertung der Wahrscheinlichkeiten liefert auch einige Erkenntnisse zu dem Zeitreihen-Prozess und den nicht exogenen Einflüssen darauf. [1,2] Die Wahrscheinlichkeiten in einem Zustand zu bleiben sind sehr hoch. Die Wahrscheinlichkeiten für Zustandswechsel sind sehr gering. Es ist daher davon auszugehen, dass $s_t$ keinem zufälligen Prozess folgt.

Einige markante Charakteristiken der vier Zustände aus der SOM können wir in den aus dem Markov-Switching Modell ermittelten Zuständen wiederfinden (siehe Abbildung 4):

– **Zustand 1: Geringe Motorlast, in der SOM '(zu) geringer Belastung' beschrieben**

|  | Zustand 1 | Zustand 2 | Zustand 3 | Zustand 4 |
|---|---|---|---|---|
| **Zustand 1** | $0,967$ | $0,0006$ | $0,005$ | $0,014$ |
| **Zustand 2** | $0,0004$ | $0,937$ | $0,016$ | $0,006$ |
| **Zustand 3** | $0,012$ | $0,051$ | $0,978$ | $0,002$ |
| **Zustand 4** | $0,021$ | $0,011$ | $0,002$ | $0,978$ |

**Tabelle 1.** Geschätzte Wechselwahrscheinlichkeiten.

– **Zustand 2: Verluste in Separierung erhöht, in der SOM mit '(zu) hoher Belastung' beschrieben**
– **Zustand 3: Durchgängig hoher Durchsatz, in der SOM als 'Gute Leistung mit hohem Durchsatz' beschrieben**
– **Zustand 4: Gemischter Durchsatz und gemischte Motorlast, in der SOM als 'gemischte Leistung mit teils höheren Verlusten' beschrieben.**

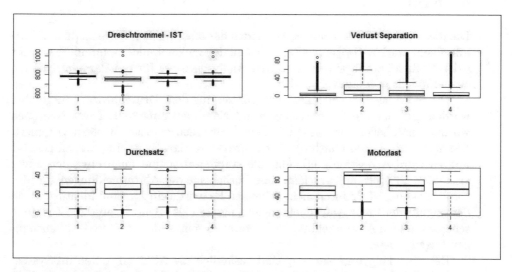

**Abb. 4.** Darstellung unterschiedlicher Variablen der Maschine je Zustand der Zeitreihenanalyse.

Unser Interesse bei diesem Ansatz galt dem Verständnis und der Vorhersage des Fehlercode bei Motorüberlastung. Der Fehler tritt auf, wenn der Motor die Nenndrehzahl nicht mehr halten kann. Kein Zustand korrespondiert eindeutig mit dem Auftreten des Fehlers, auch wenn in Zustand 3 („Maximaler Durchsatz") die Wahrscheinlichkeit, den Fehler zu beobachten, um 18% höher ist als durch die relative Verteilung der Zustände angenommen (Tabelle 2). Dieser relative Vergleich sollte aber mit Vorsicht behandelt werden, da ein detaillierter Blick auf einzelne Fehlermeldungen zeigt, dass bei auftretendem Fehlercode oft auch ein Zustandswechsel vorher und/oder nacher erfolgt. Diese Zustandswechsel sind noch näher zu untersuchen.

| | Zustand 1 | Zustand 2 | Zustand 3 | Zustand 4 |
|---|---|---|---|---|
| Anteil Gesamtbeobachtungen | $16,15\%$ | $11,56\%$ | $45,50\%$ | $26,78\%$ |
| Anteil Beobachtungen m. Fehlercode | $11,94\%$ | $10,45\%$ | $53,73\%$ | $23,88\%$ |
| Relation der Anteile | $0,74$ | $0,90$ | $1,18$ | $0,89$ |

**Tabelle 2.** Anteile der Beobachtungen je Zustand.

Ein weiterer Versuch der Verwendung des Modells für ein besseres Verständnis des Auftretens der Motorüberlastung war die Analyse der Modell-Residuen mit der Idee, dass große Modellabweichungen ein Indiz für eine Motorüberlastung und damit ein Fehler ist. Hier konnte kein Zusammenhang erkannt werden.

## 4   Fazit

Die Idee dieser Arbeiten war es, Methoden des maschinellen Lernens (hier: SOM) mit einem Markov-Switching Zeitreihenmodell zu kombinieren, um unüberwacht gelernte Zustände in einem separaten Analyseprozess für die Parametrisierung eines Modells wiederzuverwenden.

Die SOM ist auf den Mähdrescherdaten aus dem Druschprozess ausgeführt worden und hat die Beobachtungen auf Neuronen aufgeteilt. Diese Neuronen wurden anschließend manuell mit Domänenwissen zu vier Clustern gruppiert. Die derart erkannten Cluster wurden für das Markov-Switching Modell genutzt und es wurden Zustände erkannt, die stabil waren und auch etwa den zuvor erkannten Clustern entsprachen. Eine Vorhersage der Alarmmeldungen zur Motorüberlastung ist dadurch nicht gelungen. Es wurde ein Zustand identifiziert, in dem signifikant mehr Alarmmeldungen auftreten als erwartet, aber die Zustände scheinen um die Alarmmeldung nicht stabil zu sein, sodass weitere Untersuchungen folgen müssen.

Bei einer Folgeuntersuchung sind zusätzlich die Abhängigkeiten im System des Mähdreschers zu berücksichtigen. Hier wurde zur Vereinfachung der Realität auf eine Modellierung der gegenseitigen Abhängigkeiten verzichtet. Das Modell sollte aber diese Komplexität in Zukunft reflektieren.

Diese starke Vereinfachung der Realität ist ein Kritikpunkt der hier vorliegenden Untersuchung. Eine weitere Schwäche ist die geringe Datenmenge. Es wurde nur eine einzige Maschine untersucht. Eine Ausweitung auf eine größere Anzahl an Maschinen ist wichtig für verallgemeinerbare Ergebnisse. Zusätzlich sind weitere Furchtarten in die Untersuchung aufzunehmen und weitere Alarmmeldungen und deren unter Umständen vorliegenden gegenseitigen Abhängigkeiten.

Die beiden Methoden SOM und Markov-Switching Modell sind durch menschlichen Input in Form eines Domänenexpertens verknüpft worden. Der Experte hat auf Grundlage der SOM die Anzahl möglicher Zustände im Druschprozess bestimmt. Diese Anzahl wurde im Markov-Switching Modell wiederverwendet. Experteninput ist potentiell subjektiv und wenig skalierbar auf komplexere Ergebnisse unüberwachter Modellierungsmethoden. Da die definierten Zustände der SOM in den Zuständen des Markov-Switching Modells bestätigt werden, bewerten wir dieses Vorgehen hier als erfolgreiche Verknüpfung.

Insgesamt ist eine Vorhersage des Alarms überlastung des Motors im Druschprozess hier nicht gelungen. Die Ansätze sind jedoch vorhanden, diese Anomalien weiter zu untersuchen. Oben aufgeführte zukünftige Erweiterungen sind dabei zu berücksichtigen.

# Literatur

1. Hamilton, J.D.: A New Approach to the Economic Analysis of Non-stationary Time Series and the Business Cycle. Econometrica 57, 357–84 (1989)
2. Hamilton, J.D.: Analysis of Time Series Subject to Change in Regime. Journal of Econometrics 45, 39–70 (1990)
3. James, G., Witten, Daniela, Hastie, T., Tibshirani, R.: An Introduction to Statistical Learning with Applications in R. Springer, New York (2013)
4. Kohonen, T.: Self-Organizing Maps. Springer, Berlin (2001)

# Adaptable Realization of Industrial Analytics Functions on Edge-Devices using Reconfigurable Architectures

Carlos Paiz Gatica[1] and Marco Platzner[2]

[1] Weidmüller Interface GmbH & Co. KG, Klingenbergstraße 16, 32758 Detmold, Germany
carlos.paizgatica@weidmueller.com
[2] Paderborn University, Warburger Str. 100, 33098 Paderborn, Germany
platzner@upb.de

**Abstract.** Machine learning algorithms play a significant role for the realization of industrial analytics functions, such as predictive maintenance. This paper first outlines the workflow and topology variants for industrial analytics, and then focuses on the efficient realization of machine learning algorithms on edge devices using reconfigurable System-on-Chip architectures, showing the benefits of an optimized application-specific realization.

**Keywords:** Predictive Maintenance, Machine Learning, Reconfigurable System on Chip, ReconOS, k-NN, Support Vector Machine

## 1 Motivation and Application Areas

The undergoing evolution of automation technologies is driving production plants towards more flexible, but at the same time more complex production systems. These developments (e.g., vertical and horizontal integration, modularization) bring many advantages and opportunities. Nevertheless, there are also challenges implied. One such challenge is the ever-increasing need to improve plant-availability and productivity while reducing operation costs. The use of industrial analytics approaches is a promising way to take advantage of production data to tackle these challenges. The basic principle of these approaches is the use of data from production plants (e.g., sensor signals, production planning) to generate information, which can be used to improve productivity or to reduce unplanned down time.

One example of an industrial analytics function in the area of plant maintenance is anomaly detection. Here the use of data from a production system combined with industrial analytics methods and domain knowledge leads to the realization of monitoring systems able to automatically detect changes in the behavior of a machine during operation. This approach is based on the use of a normality model derived by machine learning algorithms trained with data from the production system.

Another use case is the prediction of a machine failure. Here the idea is to develop models, which can predict undesired machine states, so that maintenance can be scheduled to avoid unplanned down times. A specific challenge is to deal with unbalanced data distributions, where there are only few examples of the failures that shall be predicted. Furthermore, external and internal influencing factors and their effect on the data must be well understood. A further use case is the monitoring of production quality. Here the idea is to detect changes of the machine or in the process, which can potentially affect the quality of the final product. Application areas where industrial analytics approaches can have great impact are high-performance production plants with their typical 24/7 production schedules, such as in the automotive industry, packaging machines, or paper processing.

This paper is organized as follows: Section II presents the workflow of a typical industrial analytics function. In Section 3, topology variants for the realization of industrial analytics functions are analyzed. Section 4 focuses on realization variants on edge devices utilizing reconfigurable system-on-chip, and Section 5 presents experimental results for two machine learning techniques. Finally, Section 6 concludes the paper.

## 2    Industrial Analytics Workflow

Industrial analytics functions are typically composed of different tasks, as shown in **Figure 1**. The figure shows the typical workflow of an industrial analytics application, where data from the different devices are first consolidated in a single data source (*data storage*). The next step is to pre-process the data as preparation for the learning process (*preprocessing*). In this step, relevant features are extracted from the raw data signals, involving the combination of statistical methods with domain-knowledge to select meaningful features.

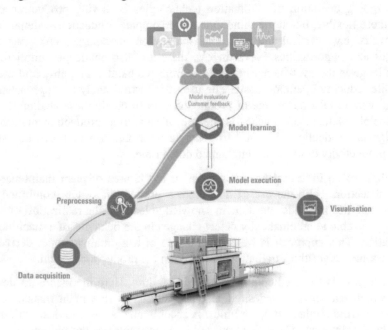

Figure 1: Typical workflow of an industrial analytics system.

The next step is the selection, training and tuning of machine learning algorithms in order to derive a model from the selected features (*model learning*). Again, the combination of analytics expertise and domain knowledge is key to develop an efficient model. Once developed, the model can be used at runtime to monitor the machine or process (*model execution*). To be useful the results need to be properly visualized (*visualization*). The kind of visualization should be selected according to the role of the person who shall use this information, e.g., the machine operator, the maintenance manager, etc. The integration of an industrial analytics function in an automation system can be done at different levels, for instance at the machine, or using a cloud platform. These possibilities are explored in the next section.

## 3     Topology Variants for the Realization of Industrial Analytics

The realization of the basic operations of an analytics function (*data storage, preprocessing, mode learning, model scoring, and visualization*) can be done at different levels of an automation system, as shown in **Figure 2**. The figure shows an example of an automation system, where data is collected at the field level using various data sources, such as remote I/Os, and an additional data collector, which, e.g., can be used to acquire energy consumption. Since data sources can be heterogeneous there is the requirement to transform the data into a unified format for further analytics processing. Depending on the given application and on the prerequisites of the user application, data can be stored either locally on premise, e.g., by using IPCs, or in the cloud using suitable storage cloud services. The analytics processing can be performed on various devices or platforms, IPCs or using software as a service in cloud platforms. While automation components are primarily used for process control, they might qualify for implementing analytics processing functionality if a suitable amount of resources is left.

Besides the given hardware resources, processing and memory requirements vary according to the task to be performed. Especially, deriving a machine model often requires significantly more resources than executing that model. There are various architectural options for storing and processing data, and the selected implementation is subject to constraints, such as the given architecture, the processing needs, data rates, and storage complexity. There is a need for flexibility in the realization of analytics functions in order to address the various industry applications. For machinery applications, data sets are generated from control systems operating in real time. The applied algorithms need to show low latency, and the data sets are typically of small volume and are highly correlated to each other. Therefore, an implementation of industrial analytics functions using edge devices brings many advantages, such as short reaction times and decreasing network traffic.

When edge devices are based on reconfigurable hardware architectures, the realization of a model can be done in a way that the hardware resources are optimally allocated depending on the given task (pre-processing, model learning, or model scoring). In the next section, different realization variants using state of the art system-on-chip architectures are presented.

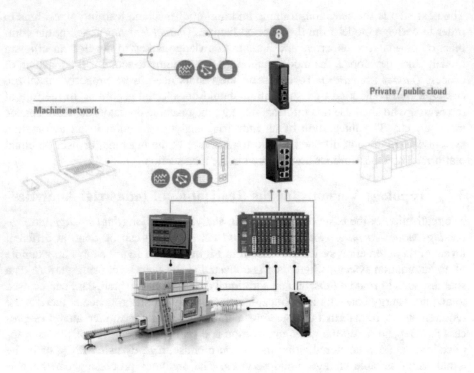

Figure 2: Example of an automation system with industry analytics functionalities.

## 4      Realization Variants at the Device Level

Modern reconfigurable architectures are system-on-chip combining CPU cores with programmable logic, memory, and peripherals. On such devices, industrial analytics functions can be implemented in software or/and in the programmable hardware for improved performance or energy efficiency. Reconfigurable system-on-chip allow for dynamic hardware reconfiguration, i.e., changing the hardware functionality at runtime. Some devices even support partial hardware reconfiguration where only a fraction of the hardware resources undergoes a reconfiguration. Such architectures support the implementation of flexible systems that can adapt their software and hardware at runtime to optimize performance and energy efficiency for the given task.

For realizing different variants of industrial analytics functions we leverage ReconOS [1], an operating system layer for reconfigurable system-on-chip. ReconOS extends a guest operating system, e.g., Linux, and provides the POSIX pthreads multithreaded programming model for software and hardware. Under this model, hardware functions are turned into so-called hardware threads that transparently communicate and synchronize with other threads and the operating system kernel.

Figure 3 shows an example for a ReconOS system with a CPU running Linux as host operating system, the programmable hardware area organized into two reconfigurable slots, memory, and further peripherals. Each reconfigurable slot can accommodate one hardware thread at a time. A hardware thread is connected to the operating system kernel on the CPU by its operating system interface (OSIF) and to the system memory through its memory interface (MEMIF). The ReconOS memory subsystem features a memory management unit that enables the use of virtual addresses in hardware threads. Internally, a hardware thread is partitioned into a state machine (OSFSM) that implements the sequential interaction with the operating system kernel and the thread's computing functionality, which typically exploits the massive low level parallelism provided by programmable hardware. For each hardware thread, there exists a delegate thread on the CPU, serving operating system requests of the corresponding hardware thread. The delegate thread concept not only simplifies the implementation of operating system services for hardware threads but also provides transparency. Software and hardware threads can communicate and synchronize without having to know whether the other threads are currently executed in software or hardware.

ReconOS supports the implementation of industrial analytics functions both at design time and at runtime. At design time, the architecture and programming model of ReconOS facilitates a step-by-step development process. First, a developer can create a software-based multithreaded prototype on a workstation, then move the application to the embedded edge device and, finally, flesh out hardware versions of threads that should be optimized. Once software and hardware versions of the application threads are available, ReconOS simplifies design space exploration, i.e., the quick creation of different hardware/software variants. At runtime, the thread management functions of ReconOS and the partial reconfigurability of the device provide the system with flexibility. ReconOS is able to dynamically load and execute software and hardware threads, and for given threads to switch between software and hardware execution to adapt to changing tasks and optimization demands.

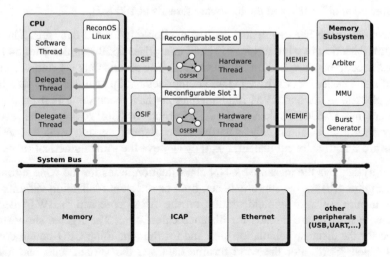

Figure 3: ReconOS architecture with two hardware threads (taken from [1]).

# 5    Experimental Results with k-NN and SVM

We present the implementation of two machine learning algorithms, the classifiers k-Nearest Neighbour (k-NN) and Support Vector Machine (SVM) [5]. The basic version of k-NN takes in a new sample and computes the Euclidian distance to all available, already classified samples with respect to the dimensions of the feature space. The majority class among the k data points with the smallest distance to the new sample is then chosen as class decision. While k-NN is a rather simple technique, it requires many runtime-intensive distance calculations to classify each new sample and sufficient memory to store all samples. SVM are algorithms for statistical learning, which in a distinct training phase compute a linear decision boundary between classes. Since typically classes are not linearly separable, SVM apply transformations of the feature space and determine separating hyper planes in a higher dimension and corresponding support vectors. In the classification phase, SVM basically computes for each new sample a single matrix-vector multiplication.

Different hardware/software versions for both k-NN and SVM have been implemented and evaluated on a Xilinx Zynq reconfigurable system-on-chip (ZedBoard) [2]. For k-NN, the ARM CPU reads the input samples and scales and assigns the resulting class decision. The most runtime-intensive part, the distance calculation, has been implemented as a hardware thread. Since the hardware thread's local memory is limited, the CPU partitions the sample data into blocks, which are consecutively transferred to the hardware thread's local memory to be processed there. For SVM, the classification step has been implemented in a hardware thread. Again, the CPU partitions the support vectors into blocks, which are consecutively transferred to the hardware thread's local memory. For both classification algorithms, the ARM CPU have been clocked at 667 MHz and the hardware threads at 100 MHz.

The classifiers have been evaluated on two available industrial data sets. The first is from fault detection in wind turbines [3] and comprises 6550 samples with 544 attributes that need to be classified into eight classes. The second data set stems from diagnosis of a sensor-less servo drive [4] and comprises 58500 samples with 48 attributes and 11 classes. k-NN and SVM implementations have been done first on a desktop computer (Intel i7-6700K), then on the ARM processor (Cortex-A9) of the reconfigurable system-on-chip, and finally as hardware-accelerated version using ReconOS. All variants are functionally equivalent, i.e., they achieve the same numerical results.

Figure 4(a) displays the measured k-NN classification times for the wind turbine data set and Figure 4(b) for the servo data set, both or different realization variants. "SW" denotes a pure software implementation on the ARM core and "nHWT" denotes a variant with n hardware threads for distance calculation. The figures show the time required for the distance calculation step and for the remaining CPU-bound computations, labeled "Sort". For the wind turbine data set, the variant with one hardware thread is actually slower than software only, but by adding more hardware threads a speedup can be achieved. For example, the variant with three hardware threads results in a speedup of 1.98 over the software solution. For the servo data set all hardware versions achieve speedups, but the CPU-bound sorting takes substantial runtime due to the large number of samples and, thus, leads to a saturation effect.

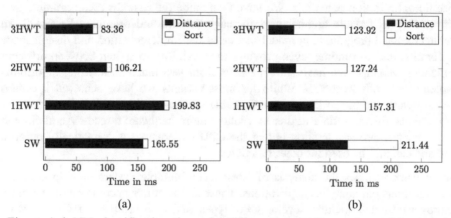

Figure 4: k-NN classification times for different realization variants: (a) wind turbine data set, (b) servo data set (taken from [2]).

Figure 5 displays the measured SVM classification times for both data sets. Again, the remaining software execution time, labeled "Decision", is small for the wind turbine data set and the overall runtime is dominated by the runtime of the hardware threads. With only two hardware threads, no acceleration has been achieved. For the servo data set all variants with hardware threads achieve speedups.

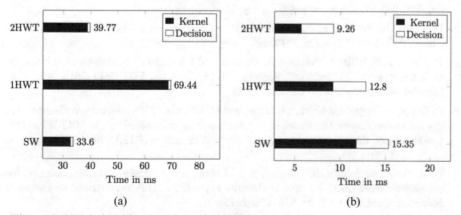

Figure 5: SVM classification times for different realization variants: (a) wind turbine data set, (b) servo data set (taken from [2]).

# 6     Conclusion

In this paper, we have outlined the workflow and topology variants for industrial analytics functions. Implementing analytics functions on edge devices can bring advantages but requires flexible computing architectures able to realize application-specific optimization potentials. We have then made the case for reconfigurable system-on-chip as flexible architectures for edge devices and presented ReconOS, an architecture and programming model that eases the implementation and design space exploration and, at runtime, even provides the flexibility to switch between different realization variants. We have shown results from two machine learning algorithms implemented with ReconOS. While for most variants we have achieved speedups over software, the use of larger reconfigurable system-on-chip would allow us to reduce the classification time further by running more hardware threads. An additional effect of hardware acceleration is that the CPU experiences a dramatically reduced load and can thus be used for other functions.

Future work includes the analysis of latest reconfigurable system-on-chip architectures for industrial analytics applications. These architectures become more and more heterogeneous and include several CPU types and even embedded GPUs. More broadly, in the emerging field of embedded machine learning the challenge is to study the trade-offs between the quality of machine learning algorithms and the performance, energy consumption, and cost when implemented on resource-constrained devices.

# References

1. A. Agne, M. Happe, A. Keller, E. Lübbers, B. Plattner, M. Platzner, and C. Plessl. "ReconOS – An Operating System Approach for Reconfigurable Computing", IEEE Micro, 34(1):60–71, IEEE Computer Society, 2014.

2. U. Riaz. "Acceleration of Industrial Analytics Functions on a Platform FPGA", Master's Thesis, Paderborn University, 2017.

3. P. Santos, L. F. Villa, A. Renones, A. Bustillo, and J. Maudes. "An svm-based solution for fault detection in wind turbines", Sensors, vol. 15, no. 3, pp. 5627–5648, 2015. Available: http://www.mdpi.com/1424-8220/15/3/5627

4. C. Bayer, O. Enge-Rosenblatt, M. Bator, and U. Moenks. "Sensorless drive diagnosis using automated feature extraction, significance ranking and reduction", in 2013 IEEE 18th Conference on Emerging Technologies Factory Automation (ETFA), Sept 2013, pp. 1–4.

5. G. A. Susto, A. Schirru, S. Pampuri, S. McLoone, and A. Beghi. "Machine learning for predictive maintenance: A multiple classifier approach", IEEE Transactions on Industrial Informatics, vol. 11, no. 3, pp. 812–820, 2015.

# The Acoustic Test System for Transmissions in the VW Group

## From Micrometers to Decibels:
## Using Real-Time Vibration Analysis and Machine-Learning

Dr. Thomas Lewien[a], Dr. Ivan Slimak[b], Dr. Pyare Püschel[a]

[a] Discom GmbH - A Brüel & Kjær Company, Neustadt 10-12, 37073 Göttingen, Germany
[b] ŠKODA AUTO a.s., Dělnická 531, 543 18 Vrchlabí, Czech Republic

**Overview**

Building automotive transmissions is a complex industrial process with high capacity and value. They transmissions are evaluated on test stands simulating vehicle-driving conditions. Using an NVH analysis of the generated vibration, production defects and vehicle noise comfort issues are identified. The presented real-time analysis provides quality parameters assessed by a combination of machine-learned and user-defined limits.

**Introduction**

The plant of ŠKODA AUTO in Vrchlabí has a long manufacturing history, which started in 1864 with the production of carriages and in 1908 with the production of cars. The story continued with the car production after 1991, when the plant joined the VW Group as one of the three plants of ŠKODA AUTO in Czech Republic.

After 20 successful years within the VW Group, there came the year 2011 and an increased customer demands worldwide with a need for a new product and production strategy. There was a need to increase the car production worldwide and in the Czech Republic as well. The plan for the three Czech plants was to concentrate the car production into two plants only, increase the overall output in those two plants and to transform the Vrchlabí plant from car production into a high-tech facility for automatic transmissions.

**Challenges of Transformation**

To be cost effective, the new production facility should deliver the automatic dual clutch transmissions in high volume to all car plants of the VW Group in Europe, South Africa and India. Our plant meant to supply not only for SKODA models, but also for the brands like Volkswagen, Audi and Seat. There were many challenges as in any automotive project, but the two most demanding were the time and the people. We had only 18 months to build new production hall, install the new machinery and step down the car production. Parallel with those activities we had to train our staff and start to

© Springer-Verlag GmbH Germany, part of Springer Nature 2020
J. Beyerer et al. (Eds.), *Machine Learning for Cyber Physical Systems*, Technologien für die intelligente Automation 11,
https://doi.org/10.1007/978-3-662-59084-3_10

produce the transmissions. This was a challenge for the people who were very good at car production, but had no experience in gear manufacturing that is so critical for the customer experience. We did not want to disappoint our customers buying cars like ŠKODA Superb, ŠKODA Octavia, VW Golf or Audi A1 and A3.

**Fig. 1.** Cars in which the ŠKODA built DQ200 automatic transmission is used

## New Demands

There was a new level in the manufacturing as far as the precision concerns. With little simplification, one can say that in the car production, the precision is in millimeters but in the production of transmission, the micrometer scale is used. Even the slightest change in micrometers can cause the increase of transmission noise by several decibels, which would annoy the customers. Therefore, several parameters of the involute with tolerances of $20\mu$ and in many cases even of $5\mu$ define the shape and the surface quality of each gear flank. For comparison, the men's hair thickness is approx. $60\mu$. Even the parts that are within the drawing tolerances can have a difference of noise level by 15dB.

Each transmission consists of several hundred parts among them 19 gears and shafts with thousands of dimensions, which must fulfill the specifications. Therefore, by daily production of 1.000 transmissions there are literally millions of parameters, which must achieve high precision at a very high cadence.

## How to Support the Production Effectively?

The repeatability of each tooth flank is essential and thus the stability of all manufacturing steps like e.g. hobbing, heat treatment, grinding, honing etc. We have been looking for a set of analytic tools, which would give our employees a constant and very specific feedback about the machines and machining tools. The automated test benches at the end of the assembly line perform testing of all customers' functionalities including noise testing for each transmission. The goal is to not only tell whether the transmissions are o.k., but also in case of noise levels to tell which part and which parameter causes the problem, e.g. input shaft 6$^{th}$ gear – tooth damaged, shifting 4$^{th}$ gear – flank shape is out of specification or ring gear – surface grinding is missing.

We are constantly comparing the data from daily test drives of our transmissions in cars from the end of line test benches, from the geometric measuring room and from the shop-floor measurement.

It enabled us very quickly to reduce the number of transmissions, which are not conforming to noise standards well below 1% and to reduce the tooling costs for by more than 10%. Moreover, the customer demand for our transmissions has risen and we have enlarged our plant to increase the daily production by 100%.

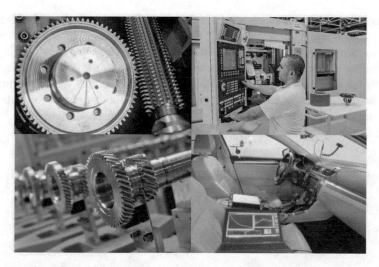

**Fig. 2.** Manufacturing of the DQ200 transmissions and the noise analysis system for vehicle correlation.

## Architecture of the Noise Analysis System

Several test stands are involved in a transmission production plant to test the assembled products. Each test stand is equipped with an NVH analyzer that measures vibration data via accelerometers that are contacting the transmission under test. The analyzers store original sensor data locally and a set of calculated result data that are handed over to a central server for storage and off-line evaluation.

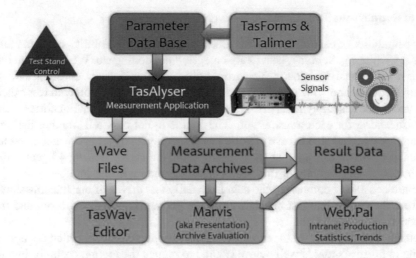

**Fig. 3.** The architecture of the NVH analysis system

### Real-Time Data Analysis

Transmissions consist of a set of gears that transform the torque and speed of the engine to the wheels. While the test stand shifts through these gears, accelerometers measure the transmission's vibration. Our real-time analysis uses the speeds of the input and output shafts to perform a data preprocessing that separates the complex transmission noise into the noise sources generated by the individual rotors inside the unit by a method called synchronous order analysis. These signals then serve for further time domain and order based spectral analysis. The resulting curves are either evaluated in their entirety or by extracting single values.

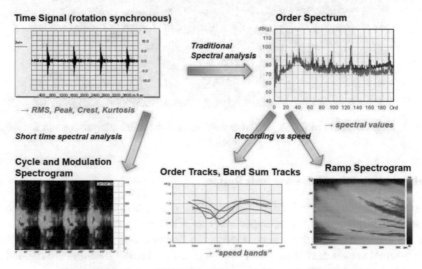

**Fig. 4.** Analysis methods of the Discom noise analysis system.

## Automatic Evaluation

The main goal of the evaluation method is to separate the good from the noisy transmissions with a high level of confidence. There are two aspects of this goal:

1) Find observable values that are specific for an individual error in the assembly and that ideally exhibit a clear separation for a known problem.
2) Automatically calculate limits by learning algorithms for most the values. Those automatically generated limits can be overridden in cases where vehicle tests indicate the necessity of fixed limits to guarantee comfort levels for the customers.

For transmission analysis, time domain Crest and Kurtosis values have shown to give good correlation to dents on individual teeth of the gears. The sensitivity can be augmented by calculating floating Kurtosis values that are evaluated in an angle domain.

To determine tooth contact geometry problems, the evaluation of gear mesh order amplitudes and their harmonics is the standard procedure. For vehicle correlation, the gear mesh order amplitude is tracked against speed and in some cases against torque to simulate the performance of the transmission in the vehicle. From these tracks, single values can be extracted that are specific to critical noise transfer paths in the final vehicle and have noise comfort relevance.

Gear defects like pitch errors, eccentricities and surface waviness are represented by spectral amplitudes at orders that identify the special fault. In addition, bearing defects can add a variety of both spectral components as well as knocking noises that are identified by modulation spectrograms

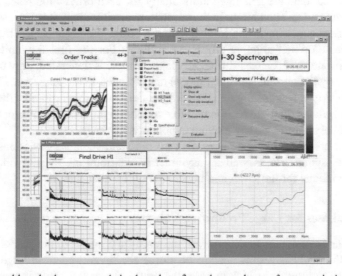

**Fig. 5.** Several hundred curves and single values form the result set of a transmission NVH test.

Altogether, about a hundred curves and several hundred single values compose the fingerprint of one transmission test. These values are then compared against individual limits. As the values are specific for certain defects, it is possible to assign error codes and messages to each value. A report consisting of these messages is then generated in case limits are violated. A priority list and a set of shadow rules are applied that consider, that follow up errors are naturally generated by certain events: A nick with a high Crest value will cause also a variety of spectral errors which are then suppressed to form a clear report.

Transmissions with errors are subjected to disassembly. The parts indicated by these machine-generated reports are scraped and the rest of the parts is fed back into the assembly process. This procedure relies on the capability of the noise analysis system to identify the root cause with a high probability.

## Generation of Limits

The limits used in the evaluation are generated based on a value specific statistical model that combines the average and standard deviation within user defined boundaries.

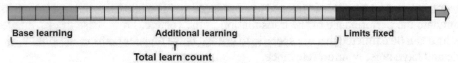

**Fig. 6.** Limit generation using statistical process parameters consists of two phases: Base learning, Additional with already calculated limits and finally fixed limits.

The statistical data are generated individually for each production test stand, transmission model and test condition. In the production environment, it is impossible to get a sample of transmissions with known status to train the algorithm. Therefore, the method relies on a self-aligning process as follows: For the first few units, the user defined limits are used. Once a sufficiently large sample is generated, only units which are accepted by the current statistical model are included in the learn set. This usually leads to a quickly self-adjusting limit set. After a defined number of units, the limits are usually fixed. As an alternative, it is possible to extend the learning of process parameters indefinitely using a slowly moving exponential average.

Evaluated process parameters consist of average and standard deviations. The system supports symmetric as well as asymmetric distributions by allowing different standard deviations for the lower and upper side of the distribution.

## Result Databases for Large Production Volumes

It is crucial to maintain an overview of the status and trends of the production plant. Therefore, the analysis results need to be stored for several years in large SQL data bases. Tools are provided which monitor the production results both in terms of TOP N rejects as well as in statistical evaluations of single values. Individual transmissions can be identified and compared to production variation. Time series analysis of single values allows the optimization of limits. Moreover, these databases potentially provide input to trend and correlation analyses ("Big Data"). The example in the left side graphic of Fig. 7 shows the time history of the gear mesh base order amplitude for 1000 transmissions. Two averages with time constants of for example 10 and 100 units are used for trend detection. The probability density function (PDF) on the right side had been fitted with two Gauss distributions for the values above and below the average for an optimized fit. The Cumulative Density Function (CDF) shows that the specified limit will reject less than 1% of the units. Algorithms for automated trend analysis are currently evaluated.

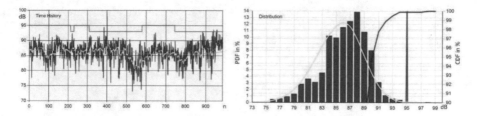

**Fig. 7.** Time history and probability density function fitted with separate Gauss curves for values above and below the average

## Verifying Limit Settings by Vehicle Correlation

Some of the measurements described above are also designed to evaluate the noise performance of the transmission in the vehicle. The limits of these measurements are usually defined by the plant's quality management as they are critical to the customer's satisfaction. They cannot be based on statistical methods and need to be evaluated by vehicle measurements. Furthermore, the vehicle acoustical results need to be correlated back to the results based on the test stand vibration measurements, sometimes leading to optimizations in the test profile. This vehicle test is supported by a dedicated noise analysis system. The result of a vehicle evaluation is shown in Fig. 8

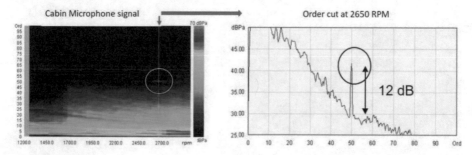

**Fig. 8.** The cabin noise shows a gear mesh whining noise at order 50 which peaks out by 12 dB from the surrounding background at 2650 RPM. This noise value is borderline and needs to be reduced by optimizing the gear geometry.

Printed in the United States
By Bookmasters